STEAM & DIESEL POWER PLANT OPERATORS EXAMINATIONS

DEDICATION PAGE

DEDICATION

To all students for their benefit and success.

Special thanks to:

The Proverbs of Solomon

Proverbs 4:7 *"Wisdom is the principal thing; therefore get wisdom: and with all thy getting get understanding."*

STEAM & DIESEL POWER PLANT OPERATORS EXAMINATIONS

Library of Congress Catalog Card No. 82-2198
Printed in the United States of America
Published by James Russell Publishing
Written by James Russell

First Edition: © July 1981, Second Revised Edition: © June 1982, Third Revised Edition: © July 1995, Fourth Edition May 2000

First 1981 edition ISBN No. 0-916367-00-2
Second 1982 edition ISBN No. 0-916367-03-7
Third 1995 edition ISBN No. 0-916367-05-3
Fourth 1997 edition ISBN No. 0-916367-08-8
Updated July 1999 ISBN No. 0-916367-08-8
Updated May 2000 ISBN No. 0-916367-08-8
e-book Edition May 2000 (No ISBN Change).

James Russell Publishing
780 Diogenes Drive, Reno, NV 89512
Phone/Fax: 775-348-8711
Email: scrnplay@powernet.net
Web site: http://www.powernet.net/~scrnnplay

THIS BOOK CAN BE PURCHASED FROM

ANY BOOKSTORE CAN ORDER YOU A COPY

SBI - Lou's Books
5647 Atlantic Ave
Long Beach, CA 90805
213-423-1403

Opamp Technical Books
1033 N. Sycamore Ave
Los Angeles, CA 9003 8
213-464-4322

- **Amazon.com**
- **Barnes&Noble.com**
- **Americanabooks.com**
- **Walden Books**
- **And other Internet bookstores.**

Chapter 1

INTRODUCTION

This test is a study guide to assist those who wish to pass Stationary Engineer and Fireman licensing exams, including Diesel Engineer examinations. The test questions and answers were initially developed to guide aspiring Engineers to pass the Los Angeles, California written and oral exams. However, these tests can be used effectively to pass Fireman and Engineer examinations nationwide, including the all important employment interviews. For well over 14 years Steam & Diesel Power Plant Operators Examinations has sold consistently. The reason being is the book works. It precisely simulates what you can expect to face when taking exams. With over 1,400 test questions and answers, it covers a vast area which leaves examiners little room to create new 'trick' exams.

The book is broken down into three main sections. 1). **500 Horsepower** - A series of questions primarily concerned with operating a small to medium size boiler plant. 2). **Unlimited Horsepower** - Questions of increasing depth which involve the operation of prime movers and larger boilers. 3). **Diesel Engines** - Questions on operating a large diesel power plant. When studying for a Fireman or Stationary Engineer exam it is advisable to take both the 500 and Unlimited exams in this book.

TERMINOLOGY TIPS

Keep in mind, every valve has a proper name in respect to its function, yet many similar constructed valves are located throughout the power plant. For example: a gate valve on a feed line may be either a stop valve (isolation) or a feed valve (control) [Preferably a globe valve should always be used for control] depending on its function. Also, the main stop valve is a gate type valve on the main steam line from the boiler so it cannot be anything but the main stop valve. To describe it as; "The gate valve on the boiler" during an exam is not specific as we know many gate valves are in use on a boiler. This will show the Inspector Examiner that you don't know what you are talking about.

Correct terminology helps on the job to communicate with Engineers and a must if you plan to advance in this field. You will find the terminology in these tests helpful to build the proper foundation to understand the deeper aspects of Engineering. There is nothing worse saying to the Examiner, "Well, I would open this valves close that one, and throttle this one over on the main pipe.

Steam Engineering is not hard to learn. Everything is basic, but just a lot of everything tied in together to make it seem confusing. Yes. It gets involved, but remember it's all basic. High pressure always flows to low pressure and high temperatures always flow to low temperatures. Find an Engineer who understands his plant and visit often. Visit many plants. (Knowing only one plant's equipment may not qualify you to pass the exam. This test covers many types of equipment found in practice.)

WRITTEN EXAM

Taking a written exam is not as easy as it appears even if it may be multiple choice as this one is. At times you will find all the answers are right, but you must pick one answer only. The key to remember is, "The answer is in the question". The question may specifically apply to a certain event only and will be seen if you read the question carefully. I have found this to be the major cause of students failing the written exam.

The oral exam will be no problem if you can communicate with the Examiner in proper terminology and most important, do you understand the steam and water cycle? Do you know what will happen even before you open or close a valve? Are you sure? When learning never accept an answer without asking, "Why is this a fact?" And never give an answer unless you can explain why! Operation procedures must be followed precisely in Power Plants. There is a reason for this. Find out why! The answer should be related to safety, never for convenience.

Some discrepancies will also appear in terminology, for example: "Water tubes will form blisters in an insulated tube's water side, but will not form a bag". Although the above example is correct, the true definition of a blister will be a laminated defect during the manufacturing of the tube. When exposed to heat and pressure this weak section will cause a small bubble and may burst. We will consider water tube localized deformation due to oil, scale etc. a blister also, but technically there is a difference. Bags occur in steam and water drum's lower section due to the presence of oil, scale, etc. in the water, and the shell is exposed to the products of combustion. Blisters occur in tubes and are only called blisters because they look like one. So therefore, blisters have a double meaning here, but you do know the difference now, I hope? Bags, of course, are large deformations whereas a blister is small.

Another example: Is an air preheater an economizer? Well, an air preheater recovers heat from stack gasses and an economizer preheats feedwater from stack gasses. Both improve the economy of fuel usage so the answer is correct. But remember that an economizer is an economizer and a air preheater is a air preheater, but they both economize. Remember to read the question carefully.

ORAL EXAM

An oral exam is a breeze when you use correct terminology. Professional Engineers love to hear the proper terms as this reveals you understand and are making serious efforts to learn. Also, the confidence comes to you because you know your not putting a fast one on the Examiner. Remember, this man is a professional and you can't fool him. "Honesty he will see, effort he will please." I've seen many try to pull wool and get sheared. Don't go for the license before you are ready. Simple as that!

My motto is: "If you think your going to pass the exam easily, you'reprobably not ready." That may sound strange but the more you learn, the more you realize what you need to learn. This is a license to learn, and you must convince the Examiner you can. This test is not a cheat sheet by any means, although many questions are likely to appear on the exam, if not most.

If you are nervous, tell the Examiner. It will be easier on yourself if you do. It sort of breaks the ice. "Don't say too much". We have all heard this time and time again, but this does not always hold true. If you cannot talk about what you know, then you're walking on thin ice anyway. On the other hand, if you talk like you know it all it will appear you are trying to make the Examiner look like he's being examined. Well, that doesn't work very well either. Tell him what you know and don't fabricate complex situations. Talk freely, but be basic and think safety.

Never argue a point with an inspector. Yes, sometimes they are wrong due to human elements and of course no man wishes to be proved so. Neither have I known an inspector to fail a person for accepting his answer. Although, arguing the point will only send you out the door with very difficult questions preceding. Last to be noted: Dress for the occasion, suit and tie are not necessary. Casual is fine, but blue jeans or work clothes are out of the question. A little respect is never out of line.

I would like to begin with "good luck", though I believe if you put in the time and effort...you won't need it.

Author
James Russell

HOW TO USE STUDY TEST GUIDE

To utilize this test as a study guide it is recommended to attend a Steam Engineering class or be tutored by an operating Engineer for best results. A steam plant text book will be very handy to find answers:. Technical book stores and college libraries will have them. More than one book is an investment. Writing manufacturers listed in the Thomas Register (located at your public library) requesting technical brochures and booklets is an option that shouldn't be overlooked. You can also write to the American Society of Mechanical Engineers' (A.S.M.E.) and the National Board of Pressure Vessels for the codes is invaluable. Do write!

When you take the following exams, read the question, find the answer, (in a text book) but never leave it at that. You must know why the other answers are wrong. After making your decision, check with the answer at the end of each chapter only after you made every effort on your own.

One sure way to make this study guide "absolutely useless" is to just look up the answer on the answer sheet attached. Anyone can do that! The goal is, "Can you think out this problem?" You will have to when taking an exam, a preliminary test to compete for a job, and most important when your on the job. Memorization is of little value in this trade. Understanding is essential. You may forget to remember what you shouldn't have forgotten. A great memory is not important and even the best forget when the heat is on. So how do you learn?

Visualize everything in your mind. If you can draw it on paper-you know it! No art work involved here. To draw the steam and water cycle only requires circles for equipment, boiler, pump, etc. and a straight line connected as pipe lines. Knowing the locations is important and direction and pressures of fluid flows. Never rush through any written exam. Go over the test three times before passing it in. First answer all the ones you know. Don't waste time on questions in doubt. Go back over the exam a second time, but this time be critical and look for trick questions. Ask yourself, "Does this question pertain to a specific event or is it just a generalization?" Then rest, clear your mind. Don't be critical on the third pass. Just go over it and ask, "Does it make sense?" The odds are the answer is correct when these three steps taken arrive at the same answer. Don't be the first to pass in the test. Look in the Table of Contents... 'Tips' are inserted in this book to help you find employment, etc.

WHERE TO FIND TEST ANSWERS AND EXPLANATIONS

After you take these multiple choice tests you will find answers to the specific chapter at the closing end of the book. The Table of Contents will guide you to the page number. Some of the answers have a [*] asterisk mark indicating a brief explanation is available. Again, the Table of Contents will guide you to the 'Explanations' page.

WHAT IS THE DIFFERENCE BETWEEN 500 HP AND UNLIMITED?

This book is written in two sections; 500 Horsepower and Unlimited Horsepower. It would be too difficult for the reader to learn all in one shot, so the book is broken down into smaller segments. Primarily, the 500 HP section is more basic and can be used for low pressure boiler exams, up to 500 horsepower examinations. It is recommended when you do prepare to take a boiler exam that you study also the Unlimited Boiler section. You just never know what questions will be asked, so be well prepared. Study the entire book for best results!

DEFINITION OF TERMS

Back Pressure Valve -Usually a weighted valve placed on exhaust line to maintain back pressure. Excess steam is often discharged to atmosphere.
Blow-Off or Blowdown Valve - Special designed straight-away valve for draining and isolation.
Boiler -Pressure vessel to change liquid to gas via heat.
Butterfly Valve -Double ported sliding gate valve. Found on receivers to control steam to feed pump.
C.B.D.V. -Continuous blow down valve controls bleeds water from boiler to control T.D.S.
Centrifugal Pump - Spins water with impellers to create pressure.

Check Valve-A simple gravity operated automatic one-way valve. Prevents reverse flow.
Closed Heater - Heating medium not in contact with fluid to be heated. Uses coils or tubes.
Condenser - Converts gas to liquid.
Deaerator -Heater under steam pressure, relieving liquid of dissolved gasses. A open type heater.
Discharge - Delivery from a machine (pumps, compressors).
ExhaustSteam - used steam from prime mover
Exhaust Steam Quality -Normally good above 15 P.S.I.g depending on heat and moisture content.
Feed Line - Feedwater Pump discharge line direct to the boiler (s).
Firetube Boiler - Heat passes through tubes while water surrounds the tubes.
Fluid -A liquid or gas. Often the term fluid is used to describe a liquid.
Gate Valve -Consisting of a moving wedge. Used for positive isolation.
Globe Valve -Controls flow. Seat & disk construction. Fluid enters under, through, and over the valve seat.
Header - Pipes or tubes connected to a common line. The header is larger than the connecting pipes or tubes.
High Pressure Steam - Steam over 15 P.S.I.g.
Injector - Uses steam forcing cones to pump water in boiler. Similar to a inspirator but has a control handle.
Inspirator -Manually operated injector pump. Used primarily for emergency feedwater induction.
Live Steam -Steam taken directly from boiler.
Low Pressure Steam -Pressure below 15 P.S.I.g.
Main Steam Line - Main line taken directly from the boiler. Has large stop valve (s) and drain line close to boiler.
M.A.W.P. -Maximum allowable working pressure. Usually stamped on pressure vessels and valves.
Medium Pressure Steam -High pressure steam reduces to a lower pressure.
Non-Return Valve -Boiler check valve on main steam line.
Prime Mover -Live steam user. A turbine, steam engine.
Plug Cock Valve - 1/4 turn operation of handle. A rotating plug with a port hole. Used for isolation.
Positive Displacement - A pump with pistons or gears. Will build pressure immediately.
P.S.I. - Pounds per square inch. A measure of force acting on a square inch surface.
P.S.I.g. -Pounds per square inch gage. Read directly from a pressure gage.
P.S.I.a. - Pounds per square inch absolute. Gage pressure plus atmospheric pressure.
Receiver - Vented tank. Collection point for condensate for condensate pump or boiler feed pump.
Relief Valve -Relieves fluids from excess pressure. Designed for liquids only, not gas.
Safety Valve -Relieves excess steam from boiler. Designed for gas service only, not liquids.
Steam -Invisible gas made through the application of heat to water.
Steam Quality - Moisture content. See Exhaust Steam Quality.
Steam Trap - Permits liquid condensate to pass through while holding back steam.
Straight-away valve - Valve you can see through when in hand and open. Used for blowdown service.
Suction -Vacuum or head pressure to a pump inlet.
T.D.S. -Total dissolved solids. Concentrations of minerals in boiler water.
Three Element Regulator -Three sensors to activate feedwater regulator. Steam flow, water flow, water level.
Viscosity -The ability of fluids to flow or resistance to flow. Oil has a higher viscosity than water.
Watertube Boiler - Water passes through tubes while heat surrounds the tubes.

TABLE OF CONTENTS

Chapter 2

Boilers 500 HP

1. Chemical test to be taken first.
a. phosphate
b. chloride
c. total dissolved solids
(d.) sulfite

2. Sample water should be
a. cold
b. hot
(c.) room temperature
d. makes no difference

3. Gauge glass breaks, first close
a. steam connection
b. drain to column
(c.) water connection
d. gauge cocks

4. Phosphate chemical
a. P03
b. P02
(c.) P04
d. PH4

5. Sulfite chemical
a. SU3
(b.) S03
c. SU4
d. S02

6. Phosphate
a. makes impurities float
(b.) keeps impurities in suspension
c. impurities settle to mud drum
d. absorbs 02

7. Sulfite
a. absorbs CO_2
b. absorbs $CO_2 + O_2$
(c.) absorbs O_2
d. impurities settle to mud drum

8. Oil inside the water tubes will cause
(a.) blisters
b. bags
c. mud drum cracking
d. nucleate boiling

9. Foaming is caused by
a. scale
(b.) chemicals
c. amines
d. hard water

10. Priming is caused by
a. cool feedwater
b. low-hard water
(c.) load swings
d. too much blowdown

11. A dry pipe is a
(a.) steam scrubber
b. steam lance
c. soot blower
d. oil line

12. A rotary cup oil burner is
a. air atomized
b. steam atomized
(c.) centrifugal and air atomized
d. centrifugal and steam atomized

13. A mechanical burner is
a. air atomized
(b.) hydraulic atomized
c. steam atomized
d. centrifugal atomized

14. Steam atomized burner will operate on
(a.) air or steam if dry
b. steam only
c. hot air only
d. air or steam if wet

15. Gauge glass water connection is
a. longer in length in total size
b. shorter in length in total size
c. the deep connection
(d.) the shallow connection
e. the same as steam connection

16. Water column acts as a
(a.) damper and sediment trap
b. anti-foaming device
c. Backflow and circulation inhibitor
d. none of the above

17. Blowing down boiler with quick and slow opening valves
a. open outside valve lst
b. open drain then outside valve
c. close drain then open outside valve
(d.) open inside valve 1st

18. When should you blow down a boiler?
a. at high load period
(b.) low load conditions
c. makes no difference
d. medium load

19. What is nucleate boiling?
a. oil in water
b. nuclear reactor units
(c.) steam film layer in tube
d. foaming

20. Oil in firetube boilers cause
a. blisters
b. cracks
(c.) bags
d. tube bowing

21. In high pressure boilers feedwater temperature must at least
a. 212°F
(b.) 120°F
c. 180°C
d. 110°F

22. A safety valve (circle 2 answers)
(a.) 'Pops' on opening
b. has a double lift with two valve seats
(c.) has a blow back feature
d. has no blow back

23. Safety valves have
a. 1 adjustment
(b.) 2 adjustments
c. 3 adjustments to close valve
d. no adjustments

24. A boiler which has its low water fuel cut-out removed
a. run only at 50% rating
b. must not be run
(c.) Cut-out device is not a mandatory requirement of the American Society of Mechanical Engineers code (ASME code).

d. must trip fuel valves at all loads

25. To temporarily stop a stay bolt leak

a. take boiler off the line
b. weld it with unit running
c. let it leak, reduce pressure 50%
d. drive a nail or pine plug in the telltale hole

26. Most likely to cause gauge glass failure

a. acid water
b thermal shock
c. low water
d. hard water

27. Some gauge glasses cannot withstand

a. pressure
b. acid water
c alkaline water
d. neutral water

28. The red line in gauge glass

a. gives support to glass
b. extends elastic limit
c makes it easy to read
d. prevents a 'lively' glass

29. A superheater - (pertaining to steam conditions)

a. raises steam pressure
b. temperature not corresponding to pressure
c. de-superheats steam
d. increases horsepower

30. Valves to be used on water column

a. globe
b. gate
c. plug cock
d. any of the above.

31. Water column whistle operates by a

a. electrode
b. float
c. steam sensor
d. none of the above.

32. The inlet pipe size to a boiler safety valve must not exceed

a. 3"
b. 5"
c. 10"
d. 2 .5"

33. Safety valves piped to a roof outlet

a. piping must be ridged to valve
b. piping welded to valve with drip pan
c. must have a drip pan and be unattached to valve
d. drip pan and piping not to exceed 10 feet tall

34. Riveted circumferential joints or girth seams are

a. butt strap joints
b lap joints
c. double butt-strap joint
d. welded single lap with angle stays

35. Repair-welded pressure vessels must be

a. boiled out
b. hydrostatic tested
c. acid cleaned
d. stress relieved with hydrostatic test following

36. Oil in boiler. What to do?

a. add more phosphate
b. blow down
c. take boiler off line
d. increase alkalinity

37. Boiler on oil firing. oil has entered boiler water. what could cause this?

a. leak in boiler tube
b. oil heater coil broke
c. oil P.s.i. too high
d. improper combustion

38. Tube fails in boiler, feed pump is on. What to do?

a. kill fire and shut off pump
b. kill fire and keep pump running
c. lower flame slowly keep pump on
d. lower flame slowly and turn off pump

39. Basic burner control

a. forced draft fan, fire eye, fuel shut off valve
b. ignition spark, pilot, fuel valve, low water cut out
c damper regulator, fire eye, fuel modulator, pressure sensor
d. none of the above

40. What boiler accessories can be connected to a water column?

a. steam gauge only
b. steam gauge, damper regulator, pressure limit switch
c. nothing can be connected to it
d. anything less than 1/2" pipe

41. When installing a water column, fittings must be

a. elbows
b. ASME code not required on piping
c. straight run of pipe no fittings
d. tees and crosses

42. How often should you blow down the column and glass?

a. once per day
b. once per alternate shift
c. once per week
d. Whenever in doubt or at least once per shift

43. How often should you blow down a boiler?

a. once per week
b. once per day
c. no less than twice per day
d. depends on total dissolved solids concentrations, steam load conditions and chloride levels, or once per shift, at low load / low fire

44. Safety valves must release

a. 55% of steam generated
b. steam to a flash tank with a drip pan installed
c. 98% of steam generated
d. all steam generated at full burner capacity

45. Furnace walls are lined with
a. asbestos fire brick
b. ceramic fire brick
c. slag fire brick
d. steel plate - no brickwork allowed by code

46. Slag on furnace walls (oil fired burners)
a. not to be removed, slag could take wall material with it.
b. scrape and hose down with hot water
c. remove slag with sandpaper wrapped around a straight edge, then hose with water
d. none of the above

47. Clean a furnace fireside of tubes with
a. brine and wire brush
b. water only
c. acid clean
d. brush and vacuum only

48. Soot blowers will
a. clean fire box
b. clean chimney stack
c. use boiler water to clean tubes
d. blow the tubes

49. A steam lance
a. automatic soot blower
b. superheater scrubber
c. hand held steam jet
d. warns of soot explosion

50. To properly clean water tubes (on water side)
a. turbine the tube
b. punch the tube with brush
c. use alkali chemicals
d. use soap and water only

51. Blow-off valves are of_______-construction
a. globe
b. ported angle globe type
c. straight away
d. any of the above

52. Main steam stop valves are of_____design
a. globe
b. ported plug cock
c. gate
d. any of the above

53. Boiler in battery with high water (3 gauges) it's best not to
a. close feed valve
b. blow down
c. raise firing level gently
d. regulate feed valve

54. Cutting boiler on line a rushing sound is heard (manually operated stop valves)
a. pressure not equalized
b. water level too low
c. burner kicked off
d. feedwater temperature too hot

55. Lining-up a boiler is a term used to explain
a. taking a boiler off the line
b. cutting-in on the line
c. adjusting the fuel mixture
d. adjusting water level

56. Feedwater enters steam and water drum in the
a. coolest section
b. hottest section
c. water wall upper header connection
d. anywhere in steam and water drum

57. Before changing a gauge glass
a. close water and steam connections/open column drain
b. close water and steam connections/open glass drain
c. close water and steam connections only
d. close water and steam connections and column stop valves

58. Lighting off the wall means
a. spraying fuel on hot fire brick
b. impossible to do "a"
c. spraying fuel on water wall tubes
d. impossible to do "c" and "a"

59. If you close the steam connection to gauge glass
a. water instantly disappears from glass
b. water quickly rises to top of glass
c. water slowly rises to top of glass
d. water slowly lowers to bottom of glass

60. If you close the water connection to gauge glass
a. water instantly disappears from glass
b. water quickly rises to top of glass
c. water slowly rises to top of glass
d. water slowly lowers to bottom of glass

61. Efficiency
a. input divided by output
b. output divided by input
c. feed water input x fuel input
d. input x output

62. If stack temperature rises with no load increase
a. forced draft fan failure
b. dirty burner tip
c. too lean fuel mixture
d. gas baffle broke

63. A flexible stay bolt

a. has a ball joint and no tell-tale hole
b. is thinner on one end
c. is made of low tensile strength steel
d. flexes by the inside fire sheet

64. A smoky fire is caused by
a. low oil pressure
b. high oil temperature
c. dirty burner
d. steam atomizing pressure is too high

65. A smoky stack is caused by
a. low oil pressure
b. high oil temperature
c. dirty burner
d. steam atomizer pressure is too high

66. Forced draft fan is located
a. in flue
b. in stack
c. in combustion chamber
d. in windbox

67. Induced draft fan is located
a. in the flue
b. in the stack
c. in combustion chamber
d. in the windbox

68. Air preheater located in the
a. brick work channels and flue
b. stack
c. 2nd pass
d. all of the above

69. Fuel dew point temperatures in stack will
a. corrode economizers
b. corrode boiler tubes
c. cause no problem
d. reduce steam flow

70. If the only feed pump available fails
a. lower firing rate - fix problem
b. shut down boiler
c. let low water cut-out extinguish fire so water won't rise and prime
d. open feed line by-pass wide open, quickly

71. Blow-off valves on a down boiler for inspection
a. should be open
b. should be chained closed
c. should be closed
d. should be chained open

72. Feedwater regulators should never
a. make constant adjustments
b. throttle
c. open wide
d. close completely

73. Steam continuously coming from blow-off tank vent indicates
a. normal operation
b. blow-off valves leaking
c. continuous blow-down valve is closed
d. new tanks always do this

74. In the steam and water loop, the highest pressure is found
a. in the boiler
b. before the check valve on feed line
c. in the feed pump suction line
d. in the main steam line

75. Gas is burned from a
a. burner tip and gun
b. rotating vane
c. stationary power plant boiler only
d. gas ring

76. High chloride in the boiler. you should
a. raise firing rate
b. lower firing rate
c. add more chemical treatment
d. blow-down

77. When installing a water column and glass the lowest visible water in glass should read
a. 2" from bottom of drum or fire sheet
b. exactly at the danger point
c. 6" from bottom of drum or fire sheet
d. none of the above

78. When starting any boiler, why do you open the steam drum vent?
a. to relieve pressure
b. relieve non-condensable gases
c. remove steam
d. to let air in

79. What is considered a proper water level?
a. 2/3 glass
b. 2 gages
c. 3/4 glass
d. 2 1/8 gages

80. Firetubes are usually_____when installed.
a. rolled
b. flared
c. rolled and beaded
d. welded and flared

81. Watertubes are mainly installed
a. rolled, flared and beaded
b. beaded and welded
c. rolled and beaded
d. rolled and flared

82. Which gage verifies a combustion chamber purge?
a. draft gauge-measured in inches of air
b. water - nanometer measured in inches of water
c. bourdon tube gauge
d. variable altimeter

83. Scale in boilers can
a. inhibit circulation and heat transfer
b. cause foaming
c. create impure steam quality
d. overheat blow-off line

84. A non-return valve is a
a. gate valve
b. angle valve
c. check valve
d. ported plug cock valve

85. A low water cut-out

a. cuts off the flame
b. cuts off the 02
c. closes fuel valve
d. activates an alarm

86. All boilers must have a water column
a. true
b. false

87. Are H.R.T.'s boilers truly horizontal?
a. yes - no pitch
b. no - 1" per 12' - pitch to rear
c. no - 1" per 12' - pitch to front

88. Angle stays in H.R.T. are located
a. in center line of tubes
b. below the tubes
c. above the tubes
d. none are required

89. If you walked in on a shift change and saw no water in boiler gage glass
a. fill with water
b. change gage glass
c. open tri-cocks, and shut down boiler if no water
d. Lower firing rate and add water, slowly

90. Modern (water-tube) boilers pertaining to staying
a. need support on both heads
b. require all welded construction
c. need no support - no flat surfaces
d. tube stays are required on heads

91. A safety valve
a. has no blow down
b. no huddling chamber or reaction flow
c. has immediate opening and blow down
d. opens slowly with a blow down

92. burning #6 oil must be
a. heated
b. heated and atomized
c. steam atomized
d. air atomized with forced draft

93. Water level in gauge glass (boiler operating)
a. is actually higher than in drum
b. is actually lower than in drum
c. shows exact level
d. none of the above

94. First thing to do when talking over a shift
a. try gauge cocks
b. Blow down low water cut-out and test alarm
c. blow down water column
d. blow down gage glass
e. all of the above

95. A water column has
a. 3 gauge cocks and drain
b. 2 tri-cocks and whistle
c. Low water cut-out, four gauge cocks and a drain line
d. Three gage cocks and no drain line smaller than two inches

96. Which connection is blown first on s water column and glass?
a. glass connection
b. steam connection
c. column connection
d. low water cut-off

97. Flame detector verifies
a. visible light
b. visible flame only
c. ultraviolet and / or infrared energy
d. incomplete combustion

98. A feedwater regulator regulates
a. pressure
b. pressure drop
c. flow
d. deareator level

99. Efficient combustion is verified by a
a. draft gauge
b. orsat device
c. damper controller
d. 'U' tube manometer

100. A blow-off connection exposed to a heat source is
a. bad for efficiency
b. hazardous
c. is okay if pad is welded and not riveted
d. is okay if tandem blow-off valves are installed

101. A continuous blow-down valve controls
a. total dissolved boiler chemicals
b. TDS = total dissolved silica
c. impurities in solution
d. flakes of calcium- carbonate scale particles and chloride salts

102. Feed by-pass valves are usually
a. Y' type angle valves
b. straight-away plug cock or gate valves
c. globe valves with double seats
d. globe valves
e. ported seatless valve

103. Three element feedwater regulator will sense
a. drum water level
b. drum level, feed water flow, steam flow
c. drum level, feed water pressure, steam pressure
d. all of the above

104. Welding on pressure vessels must be performed by a ______ welder.
a. certified
b. certified with OSHR classification
c. any welder under any classification or license
d. any pressure vessel ASME certified welder with stamps or seal
e. any ASME licensed engineer or boiler inspector

105. A high pressure boiler converted to low pressure

a. a larger burner must be installed
b. a taller chimney or stack installed
c. larger safety valves are to be installed
d. small burner tip installed with induced draft

106. When taking a boiler off the line, when should main steam stop valves be closed?
a. when boiler is cooled down to room temperature
b. when boiler acts as a condenser
c. right away
d. at line pressure

107. Blowing down a boiler before placing boiler on the line
a. should not be done at all
b. performed only when high water is present
c. is important for circulation and maintain even temperatures
d. causes harm to boiler

108. Cutting a boiler in battery using hand operated stop valves open 1st valve when
a. 10-15 p.s.i. of line pressure
b. 5-10 p.s.i. of line pressure
c. 5-15 p.s.i. of line pressure
d. when steam gauge reads above line pressure 30 p.s.i.

109. A superheater is a
a. Bank of standard boiler tubes on steam line in radiant section
b. heat exchanger
c. pressure and heat booster
d. an endless continuous steam loop that raises steam temperature but not pressure

110. ASME
a. American Society of Mechanics Engineers
b. American Society of Mechanical Engines
c. American Society of Mechanical Engineers
d. American Society of Master Electrical Engineers

111. Which boiler delivers highest quality steam?
a. HRT
b. watertube
c. up-right vertical firetube
d. all the same quality

112. Telltale hole pertains to
a. stay tube
b. angle stay
c. water leg stay
d. threaded through stay

113. Caulking a riveted joint refers to
a. lap and but strap joints
b. rolled tube joints
c. welded joints
d. expanding rivet in hole

114. You step into boiler room, see fire tripped out on low water and no water in the glass. You should
a. open feed valve wide open
b. close stop valves and open feed line
c. run 2 feed pumps and open bypass to regulator
d. do nothing
e. lift safety valves by hand

115. Turbulence in a combustion chamber
a. upsets draft
b. waste fuel
c. causes brick work damage
d. is desirable

116. High alkaline water in boiler causes
a. corrosion
b. caustic embrittlement
c. pitting
d. fire cracks

117. Riveted boilers longitudinal seams are
a. lap joints
b. butt-strap joints
c. stay bolted
d. stayed

18. Steam load changes will
a. affect steam p.s.i.
b. affect steam p.s.i. and water level
c. affect burners only
d. only affect prime movers

119. To free one gauge glass connection from blockage
a. blow column
b. blow glass
c. open tri-cocks
d. crossblow the glass connection

120. A non-return valve
a. protects main steam line
b. protects prime mover
c. protects boiler
d. is used only for automatic cut-in

121. A desuperheater will
a. reduce pressure
b. introduce water to steam
c. extract damaging moisture from steam
d. detour steam into the last pass of desuperheater tubes

122. Primary air will assist
a. atomization
b. combustion
c. soot blowing
d. flue gas cooling

123. Secondary air will assist
a. atomization
b. combustion
c. soot blowing
d. flue gas cooling

124. Passes of gas in a brick built HRT boiler
a. one
b. two
c. three
d. four

125. What type of valve is used on blowdown lines?
a. special design globe
b. gate with positive seal
c. straight-away

d. plug cock quick opening with extended handle with slow opening valve attached

126. Instead of forced draft fans, use

a. primary forced air steam jets
b. vacuum pumps
c. air pressure atomizers
d. a tall chimney

127. When operating steam soot blowers, first

a. open exit draft damper
b. increase forced draft
c. raise water level
d. increase steam pressure

128. Boiler with a crown sheet

a. HRT
b. watertube with box headers
c. Manning or vertical
d. Scotch Marine

129. A corrugated fire sheet needs no

a. joints at ends
b. stays
c. water to keep cool
d. hand holes for access

130. A water trough in a boiler is for

a. feed water dispersion
b. steam scrubbing
c. keeping burners cool
d. the mud drum sediment trap

131. Priming can be caused by

a. decrease in load and excessive water treatment chemicals
b. increase in load and excessive water treatment chemicals
c. load has no effect on water level
d. increase in load is the only factor

132. Foaming is caused by

a. decrease in load and too many boiler compounds
b. increase in load and too many boiler compounds
c. load has no effect on water level
d. excess boiler chemicals

133. Forcing the fires to maximum can cause

a. foaming
b. burner failure
c. tube failure
d. none of the above

134. To place a cold boiler on line, how long should it take?

a. as long as it takes on low fire to raise steam
b. 2 to 3 hours
c. 6 to 8 hours
d. watertube boilers can be brought up fast

135. If you see steam continuously leaking from blow off tank vent

a. boiler tube failure
b. blow-off valves leaking
c. no vent on tank
d. none of the above - it is normal to see steam

136. Priming is destructive to

a. high water alarm
b. superheater
c. economizer
d. air preheaters

137. The inside blow-off valve is the (quick and slow opening type)

a. quick opening non-return check valve
b. slow opening valve
c. blowing valve
d. sealing valve

138. Firetubes over 4" are called

a. main tubes
b. flues
c. air preheaters
d. none other than tubes

139. Tri-cocks are_____type valves

a. gate
b. ball cock
c. plug cock
d. globe

140. To clean firetubes for inspection you should

a. blow with water
b. blow with brine
c. turbine the tube
d. punch the tube

141. Blowing tubes. Which method is best?

a. burner at low fire
b. open all dampers - burners on high fire
c. open flue damper - draft fans at low speed
d. kill fire - open dampers and blow tubes hard

142. Tube failure can be caused by

a. too much sulfite
b. forcing the fires
c. high water level
d. steam pressure at maximum allowable working pressure

143. Blow the Low Water Cut Out (LWCO)

a. once per day
b. per shift
c. monthly
d. yearly

144. Safety valve capacity test

a. set pressure of boiler above popping point then lift handle
b. lift handle and observe blow-back
c. close stop valves - run fire to full capacity
d. partly close stop valve and run fire at 50% rating till valve "pops" and record blow-down

145. Adjustments on safety valves

a. spring tension only
b. spring and blow-back
c. blow-back only
d. no adjustments

146. Raising the lower blowback ring on a safety valve will

a. increase set pressure
b. decrease set pressure
c. increase blow down
d. decrease blow down

147. Spring on a safety valve is (drum valve)
a. round
b. square
c. double counter wound
d. octagon

148. You enter plant seeing no water in gage glass. You should
a. kill fire and do nothing to disturb feed water flow
b. kill fire - open feed valves wide
c. kill fire and feed pump
d. kill fire and lift safety valve lever immediately.

149. Two boilers in battery. One has high water the other low water
a. increase fire on boiler with high water. Decrease fire on low
b. close feed valve on high water boiler, open feed valve on low water boiler
c. shut down low water boiler and increase fuel pressure on high water boiler, then blow down
d. shut down both

150. Induced draft fan is located in the
a. windbox
b. furnace
c. chimney stack
d. flue

151. Longitudinal riveted joints are
a. butt strap joints
b. lap joints
c. double butt lap joints
d. welded butt with angle stays

152. Superheater located in the first pass of a furnace is
a. conduction type
b. flue type
c. radiant type
d. convection type

153. Most steam is produced in the
a. generating tube bank
b. water walls
c. steam and water drum
d. economizer of watertube boilers

154. When laying-up a boiler never have
a. pressure inside
b. water inside
c. steam blanket
d. vacuum inside

155. Primary superheater is located in the
a. lst pass
b. 3rd pass
c. 2nd pass
d. flue

156. Secondary superheater is located in the
a. lst pass
b. 2nd pass
c. 3rd pass
d. flue

157. Rivets in double shear are twice as strong as rivets in single shear
a. true
b. false

156. A joint which forms a true circle
a. single riveted lap joint
b. double riveted lap joint
c. any type butt joint
d. 'b' and 'c'

159. You keep raising the firing rate to maintain steam pressure, but there is no increased steam demand
a. low water in boiler
b. feed pump stopped
c. broken gas baffle
d. 'a' and 'b'

160. Bags can occur in water tube boilers if (oil in water)
a. underside of shell is insulated
b. underside of shell is not insulated
c. underside of shell is oiled
d. underside of shell is dry

161. Corrosion in boilers is chiefly caused by
a. $C0_2$
b. H_20
c. 0_2
d. alkaline water

162. Before starting a steam soot blower, first
a. open furnace explosion doors
b. open drains
c. open steam supply valve
d. raise boiler water level

163. Proper boiler water P.H.
a. 9-11
b. 5-7
c. 1-4
d. 14-18

164. A good C02 reading on a gas fired boiler
a. 12 to 14%
b. 25%
c. 100%
d. 75%

165. Heat is transferred in a boiler furnace by
a. convection
b. conduction
c. radiation
d. all of the above

166. How would you lite off a boiler with 5 burners?
a. lite each burner off of an existing burner
b. lite all of them at the same time
c. lite each burner separately

167. The air register is
a. a flue damper
b. to control air for combustion
c. used to test flue gas content
d. the path fuel travels to burners

168. Name a boiler with circulating tubes
a. cross drum with headers
b. HRT
c. vertical
d. locomotive

169. Dirty tubes in boilers
a. reduce efficiency
b. cause explosions
c. create thermal stresses
d. all of the above

170. Boiler foaming will cause
a. rapid fluctuating of water in gauge glass
b. bubbles in gauge glass
c. carryover or priming
d. all of the above

171. To cure a boiler from foaming
a. close stop valve
b. blow down and add feed water repeatedly
c. lower firing rate
d. all of the above

172. What would you do if safety valve did not open with pressure rising?
a. kill fires
b. pop valve by hand
c. check steam gauge calibration
d. all of the above in this order

173. What is a buck stay?
a. angle stay for HRT boiler
b. metal support for brick work built-up
c. through stay for HRT boilers
d. none of the above

174. Riveted lap joints are in single shear
a. true
b. false

175. Butt double-strap joints are in double shear
a. true
b. false

176. You blow down a boiler to prevent a condition called slagging
a. true
b. false

177. Why can boiler tubes be thinner than shells?
a. no joints or seams
b. reduced surface area
c. made to a true circle dimension
d. all of the above

178. Corrosion in flue is most severe in
a. radiant section
b. economizer section
c. air preheater section
d. 'b' and 'c'

179. Boilers are mainly classified as
a. stationary
b. marine
c. locomotive
d. all of the above
e. none of the above

180. Circulation of water in a watertube boiler without pumps is caused by
a. steam forming in tubes
b. gas baffles
c. heat of radiation
e. molecular anti-gravity

181. Boiler's low water alarm activates. What should you do?
a. kill the fires
b. increase feedwater
c. open drum vent and lower the fires
d. relieve all the pressure in boiler

182. Oxygen in a boiler causes pitting of the metal
a. true
b. false

183. Less fuel is used with excess air at burner
a. true
b. false

184. If load is decreased suddenly
a. steam p.s.i. will rise
b. water level drops
c. 'a' and 'b' on any boiler
d. none of the above on firetube boilers

185. Gas to be found in flue or smoke stack
a. carbon monoxide and sulfur
b. nitrogen and carbon dioxide
c. oxygen
d. all of the above
e. ammonia chloride and 'd'

186. To verify excess draft
a. read a draft gage
b. observe flame color
c. check for sparks in flame
d. use an orsat test
e. any of the above

187. Flue gas sample results should be
a. low carbon monoxide
b. low oxygen
c. moderate carbon dioxide
d. 'a' and 'b' and 'c'

188. Cooling a boiler down is not as critical as firing it up to line pressure in relation to thermal stress, therefore perform the latter much more slowly
a. true
b. false

189. A gauge glass properly installed will be in effect floating and not touching any metal
a. true
b. false

190. How can you tell a solid staybolt is broke on a boiler under pressure and on line?
a. tap with a hammer
b. remove and inspect
c. drill a tell-tale hole

191. Four boilers connected to one stack. The boiler furthest from the stack will have more draft, therefore, dampers will be closed down more than the nearest boiler
a. true
b. false

Chapter 3

Auxiliaries 500 HP

1. A reducing valve
a. reduces steam flow
b. drops p.s.i.
c. reduces steam volume
d. regulates steam velocity

2. The white cloud you see from a deareator vent is
a. steam
b. steam and air
c. steam and vapor
d. water vapor

3. A relief valve
a. pops on opening
b. lifts according to PSI rise
c. has no lifting mechanism
d. pops and blows back

4. Intercoolers are used on
a. steam turbines

b. boilers
c. gas compressors
d. oil heaters
e. all of the above

5. A valve that water enters under the seat, passes through and over the seat
a. globe
b. gate
c. plug cock
d. needle valve

6. Cyclone steam separators are used mainly on
a. steam mains
b. exhaust lines
c. steam drum outlet
d. turbine exhaust

7. A condenser's function converts
a. gas to vapor
b. vapor to heavy gas
c. steam to vapor
d. gas to liquid

8. #6 fuel oil must be heated. Another trade name for this oil is
a. fuel #2
b. Bunker - C
c. refined oil
d. # 6 low viscosity

9. Butterfly valve on a receiver (for makeup water)
a. is a double ported valve
b. single ported valve
c. globe valve
d. gate with orifice in bonnet

10. Inside a deareator
a. air ejectors and drip pans
b. steam coils
c. oil separators
d. trays, drip pans, nozzles

11. Cooling towers function on which basic principle?
a. heat transfer in reverse-cool to hot
b. air pressure
c. boiling refrigerant
d. evaporation

12. A typical outside screw and yoke valve (OS&Y)
a. feedwater regulator
b. safety valve
c. blowdown valve
d. stop valve

***13.* Zeolite is used in**
a. boilers
b. water softeners
c. water filtration tanks
d. condensate returns
e. anion reducing tanks

14. Zeolite is a
a. ion exchange mineral
b. makeup water rust inhibitor
c. makeup water turbidity filter and iron softener
d. 'c' and explosive, acidic granular spheres of pellets to be handled with caution

15. When installing a inverted bucket trap install on the inlet line
a. a drip leg
b. strainer
c. check valve
d. drip leg and strainer

16. A simple mercury switch feedwater regulator operates by
a. expansion tube
b. float
c. 3 elements
d. 2 elements
e. 'a' and 'b' and 'd'

17. Back pressure valves are found on
a. exhaust line of steam engine and steam driven reciprocating pumps
b. exhaust line of steam engine only when condensing
c. steam duplex pumps only
d. discharge line on pumps

18. A closed feedwater heater is located (in relation to boiler feed pump)
a. on suction line
b. discharge line
c. in boiler flue
d. no such heater

19. A closed feedwater heater
a. has coils that water flows through
b. trays and drip pans
c. water and steam do mix
d. heats water and removes oxygen

20. Constantly discharging trap on a closed heater indicates
a. back pressure valve did not close
b. back pressure valve stuck closed
c. broken internal coil or tube
d. wet steam from exhaust steam source
e. leaking check valve

21. A deareator extracts non-condensable gasses by
a. steam
b. heat
c. vacuum
d. pressure
e. 'a' and 'd'

22. Installing a bourdon tube pressure gauge on boiler requires
a. heavy wall pipe
b. pig tail loop only
c. water leg with test valve
d. ASME and National Board pressure vessel stamp

23. Deareator is located in relation to pump
a. pump discharge side at least 5-feet higher
b. pump suction side at least 5-feet higher
c. pump suction side at least 10-feet higher
d. pump suction side at least 20-feet higher

24. Water is pumped ________ a closed heater
a. from
b. through
c. both 'a' and 'b'
d. none of the above

25. Water is pumped ________ an open heater
a. from
b. through
c. both 'a' and 'b'

d. none of the above

26. A receiver is a
a. heater
b. storage tank
c. flash tank
d. all of the above

27. Best trap to use on fuel oil heaters
a. bucket
b. inverted bucket
c. float
d. impulse

28. Thermal feedwater regulator operates by a
a. gas or oil burner
b. expansion tube
c. float
d. steam flow recorder

29. A blowdown tank
a. is an expansion chamber
b. a flash tank
c. a sewer holding tank
d. backflow device

30. Regenerative air preheater
a. rotates in smoke stack
b. rotates in flue
c. rotates in last pass
d. does not rotate in stack

31. Air preheaters
a. conserves water usage
b. increases fuel temperature
c. saves furnace brickwork from thermal shock damage
d. saves fuel cost

32. Feedwater regulator valves have a ___________ valve configuration?
a. gate
b. ball cock
c. globe
d. balanced globe

33. First phase of regenerating ion-exchange water softener
a. rinse
b. brine rinse
c. backwash
d. all 3 at once

34. Air injectors are use
a. vacuum lines
b. condensers
c. air compressors
d. none of the above

35. Economizers
a. heat flue gas
b. increase fuel efficiency
c. save feedwater
d. pre-heat primary air

36. Receivers must be
a. tilted to the feed pumps
b. installed with safety valve
c. vented to the atmosphere
d. installed at least 10-feet higher than feed pumps

37. Thermocouple
a. measures flue gas CO_2
b. flame detector for fuel valve operation
c. measures elongation of boiler steam drum metal under stress
d. senses temperature and relays to chart recorders

38. Thermostatic trap blowing steam
a. bellows cracked
b. valve seat dirty or worn
c. bellows screw out of adjustment
d. ‘a’ and ‘b’
e. vacuum in pipeline downstream of trap

39. Compound steam gauge will measure a vacuum or *pressure*
a. true
b. false

40. Worst thing you can do to ruin a globe or gate valve.
a. open valve quickly
b. back off valve 1/2 turn when valve is open
c. use graphite on valve stem
d. never exercise valve

41. A globe valve can be installed in reverse to eliminate a noise problem, but you should check the ASME and National Board codes before doing so.
a. true
b. false

42. A deareator is vented to the atmosphere, is under pressure, and heats water above corresponding atmospheric temperature.
a. true
b. false

43. Demineralizer placed in the condensate return system will remove minerals picked up from equipment and piping in the closed steam and water cycle and is commonly referred to as a ‘polisher’
a. true
b. false

44. Carbon rings in a steam turbine are cooled by
a. air
b. steam
c. water
d. oil

45. Vacuum return pump fails. You will need to bypass pump and increase steam pressure in the heating system.
a. true
b. false

46. Steam has a cooling effect on the blades in a turbine as steam is flowing. So, at start-up, flow must be established to avoid overheating and the exhaust temperature recorded.
a. true
b. false

47. Check the proper direction of rotation when starting a turbine for it may be rotating in reverse.
a. true
b. false
c. True - if casing drains are closed turbine shaft will always rotate in reverse.

48. Non-condensable gasses entering from the feedwater and leakage on condenser joints tend to break the vacuum.
a. true
b. false

49. The condenser is larger than the evaporator in a typical refrigeration system.
a. true
b. false

50. A two-stage air ejector is________ operated and found on turbines' condenser to maintain a vacuum.

a. water
b. steam
c. air
d. vacuum

51. A safety valve that 'chatters' when closing needs a spring adjustment.

a. true
b. false

52. Some HRT firetube boilers have tube soot blowers.

a. true
b. false

53. Heat is applied by the compressor to a refrigerant gas and condensed to a liquid in the condenser by removing heat.

a. true
b. false

54. Cooling is accomplished in the______ in a refrigeration system due to heat transfer via evaporation.

a. condenser
b. evaporator
c. compressor
d. receiver

55. Pressure of a gage reads 100 p.s.i. at sea level. The absolute pressure will be

a. 100 p.s.i.
b. 85.3 p.s.i.
c. 114.7 p.s.i.
d. 86.7 p.s.i.

56. Rotary pump is best suited for pumping water.

a. true
b. false

Chapter 4

Compressors 500 HP

1. Some reciprocating compressors are unloaded by

a. relief valves
b. finger valves
c. plug cock valves
d. gate valves

2. Centrifugal compressor has

a. vanes and impeller
b. pistons
c. rotary piston
d. poppet valves

3. Reciprocating compressors are started

a. loaded
b. unloaded
c. intercooler off
d. vanes in low position

4. Intercoolers are located on compressors

a. after last stage
b. between any stage
c. before the first stage
d. inside the compressor cylinder

5. Compressed air contains (reciprocating 4 stages-no coolers)

a. pure air
b. air and steam vapor
c. air, oil, steam
d. air and oil

6. Second stage in any compressor is

a. larger
b. smaller
c.. same size
d. water jacketed to cool oil

7. Which compressor requires unloader valve gear?

a. centrifugal
b. piston reciprocating
c. rotary screw
d. turbine

8. Aftercoolers on compressors are located

a. after first stage
b. after any stage
c. after second cylinder compressor stage
d. after last stage

9. Aftercoolers on compressors.

a. remove oil
b. remove air
c. remove condensate
d. remove superheated air

10. Reciprocating engine or compressor has

a. vanes
b. pistons
c. **impellers**
d. **gears**

11. Any standard automotive oil can be used in any compressor

a. true - 30 weight oil only
b. false
c. 'true' if 10 weight SAE certified oil is used

12. Knocking noise from a reciprocating compressor could be a

a. wornbearing
b. water in cylinder
c. piston 'slap'
d. loose wrist pin
e. any of the above

13. Unloader failure will destroy electric drive motor.

a. true
b. false

14. Oil in high pressure air lines can promote a fire and you may not know it before it is too late.

a. true
b. false

15. Water in air lines can cause water hammer and damage to equipment and personnel.

a. true
b. false

16. Before starting a compressor check

a. discharge valve open
b. inlet valve open
c. safety valve
d. oil level
e. all of the above

17. To raise air line pressure, you should first check

a. receiver MAWP
b. compressor ratings on name plate
c. line material strength
d. lube oil flash point and all of the above

18. Compressor that requires the largest relief valve relative to line pressure.

a. 100 p.s.i.
b. 1,000 p.s.i.
c. 3,000 p.s.i.
d. all require same size

19. Compressed air is used to drive a machine requiring oil for lubrication, such as impact hand tool turbine drive. Which compressor is best suited for this application?

a. reciprocating
b. centrifugal
c. rotary vane or screw
d. axial flow

20. Reciprocating compressors need no cams to operate valves.

a. true
b. false

21. A manometer is used to measure air flow read on a scale of Cubit Feet per Minute.

a. true
b. false

22. Unlike an internal combustion engine a reciprocating compressor's oil is not contaminated by combustion blow-by products. Select one answer as to why the oil needs no changing

a. condensation in oil will not corrode metal or change oil P.H.
b. dust particles in air will not pass the intake filter
c. synthetic oil (if used) never needs to be drained
d. the oil is circulating in a sealed system, impurities cannot make contact
e. none of the above

23. A reciprocating compressor with a broken leaking inlet valve will overload an electric motor when starting.

a. true
b. false

24. Reciprocating compressor losing power.

a. worn valves
b. worn pistons and rings
c. dirty air filter
d. all of the above

25. If air is pulsating in and out of the air inlet line to a reciprocating compressor continuously

a. worn piston rings, oil blow-by increasing compression
b. unloader stuck, opening inlet valve on lst stage cylinder
c. leaky inlet valve on lst stage cylinder
d. check valve on air inlet line leaking from receiver
e. 'b' or 'c' only
f. all of the above would cause this

26. If you reversed the air flow direction in a compressor / receiver air system, you would have__________in the system

a. pressure
b. vacuum
c. oil
d. superheated air and 'a'

27. A rotary screw compressor started to make a rubbing noise. What will you do?

a. close down on suction valve
b. inject oil in suction air line
c. compressor overloaded- perform 'a'
d. open relief valve and perform 'c'

28. What will you do if the compressor's air receiver pressure is rising past the MAWP?

Write your answer to question #28 here

____
____
____
____
____
____
____
____
____.

29. A rotary screw compressor's oil separator on discharge line returns the oil to

a. sewer or waste barrel
b. compressor
c. crankshaft bearings only
d. aftercooler and intercooler to cool air
e. no oil separator on outlet.

***30.* Centrifugal compressor making a rubbing noise.**

a. oil in air
b. rotor warped
c. water in air
d. worn rotor bearing
a. any of the above

***31.* How did oil get into the air in question #30?**

a. aftercooler or intercooler tubes or coil leaking air
b. rotor bearing seal damaged
c. write your answer here__________
____
____
__.
d. none of the above

32. Centrifugal compressors are unloaded by restricting air inlet or utilizing intercooled recirculation.

a. true
b. false

33. Oil is seen discharging from an intercooler trap on a 5-stage centrifugal compressor. Could the last stage compressor seal leaking along rotor cause this if machine is running at 20% rating?

a. yes
b. no

34. give a reason for your answer to question #33.

____
____
____
____
________.

35. two reciprocating compressors taking suction from a common inlet line. #1 is running, #2 is off. As you check compressor oil sight glass you noticed the oil in #2 was overfilled to the top of the sight glass and all of a sudden the oil disappeared. What caused this to happen?

Write your answer here___________

____

____

____

____

____

36. Relief valves on a air receiver should be designed to relieve all air generated from compressor at full CFM capacities.

a. true
b. false

Chapter 5

Mechanics 500 HP

1. A nipple is a pipe

a. over 7" in length
b. under 2" in length
c. under 6" in length
d. between 21, and 7" in length

2. A helical gear is of a

a. herring bone design
b. straight cut design
c. worm cut design
d. square cut design

3. A foot valve is most likely to be found on a

a. feed pump
b. oil pump
c. condensate pump
d. condenser pump

4. It is okay to work on energized electrical panels with a wet floor when

a. one fuse is removed in a 3-phase system
b. insulated shoes are worn
c. floor is damp and no puddles exist
d. none of the above

5. To pack a stuffing box it is best to ___________ the packing around the valve stem.

a. coil
b. knot tie
c. braid
d. ring and gap-step

6. To test a soldered copper pipe line before turning on the fluid, pressurize with

a. nitrogen
b. carbon dioxide
c. oxygen
d. 'a' and 'b' mixture

7. Teflon tape should not be used on

a. steam lines
b. hot water lines
c. boiler chemical lines
d. none of the above

8. Antiseize compound is used on

a. pressed fittings
b. pressed bearing sleeves
c. flange bolt threads
d. all of the above

9. Solder used to make small electrical connections

a. 50/50 or 40/60
b. acid core
c. rosin core
d. silver solder

10. A machine thread is constant whereas a pipe thread is tapered.

a. true
b. false

11. When removing a pressed keyed piece to a shaft it is best to________when using a puller

a. cool the shaft
b. heat the keyed piece
c. apply even pressure
d. tap puller spindle with hammer
e. all of the above in this order

12. When aligning a pump and motor with a flexible coupling

a. bolt down motor then line-up pump to motor
b. bolt down pump and assemble pipe lines then lineup motor to pump
c. bolt down pump and motor, align, then make pipe connections
d. flexible couplings require no alignment

13. After lining-up a generator to a turbine. It is best to

a. taper ream and hammer home steel tapered drifts to secure generator to foundation
b. shim generator and torque bolts on turbine
c. disassemble coupling and check coupling face readings with dial micrometer
d. disassemble coupling and check coupling rim readings with dial micrometer

14. Hydraulic fluid systems operating at ambient temperatures need no pig tail or water leg when installing a pressure gauge.

a. true
b. false

15. A gear reducer drive system requires 90 weight grease. You would likely have to

a. use a grease gun
b. heat grease and pour into ceasing
c. heat casing first, then use grease gun

16. Roller bearing on a electric motor starts making noise

a. remove relief plug
b. grease fitting to bearing
c. stop motor immediately then perform 'a' and 'b'
d. 'a' and 'b'

17. Motor runs hot, temperature rapidly rising

a. use a fan to cool motor
b. grease bearings
c. stop motor immediately
d. wipe case with a wet rag
e. 'a' and 'b' and 'd'

18. When tightening electrical connection lugs
a. isolate power
b. test for voltage
c. use a torque wrench on lugs
d. 'a' - 'c' - 'b'
e. 'a' - 'b' - 'c'
f. 'c' - lubricate lug threads

19. To remove a slightly rounded nut you should next try a
a. 6 point socket
b. 8 point socket
c. nut splitter
d. heat with a cutting torch

20. OS&Y (Outside Screw and Yoke) gate valve can be installed in any position.
a. true
b. false

21. Most important valve to install on a oil heater steam line.
a. check valve
b. steam pressure regulating valve
c. safety valve
d. back pressure or reducing valve
e. oil separator

22. Valve to be used on a 'surface-blow' line.
a. gate
b. plug cock
c. globe
d. ball valve

23. Valve to be used on any blow down line.
a. gate or straight-away type valve
b. quick opening plug cock valve
c. globe valve
d. any two inch valve is okay

24. Valve to be used on a natural gas line for isolation.
a. gate
b. plug cock
c. globe
d. OS and Y with a check valve

25. Valve to be used for boiler main stop valves.
a. OS and Y gate valve
b. OS and Y globe valve
c. OS and Y ball valve
d. Any non-rising stem valve

26. Never install insulation on a steam trap for proper operation.
a. true
b. false

27. A steam trap designed for high condensate and non-condensable gas flow capacities.
a. bucket or expansion trap
b. inverted bucket trap
c. float trap
d. thermostatic float trap
e. impulse trap

28. Your plant has a small 13 p.s.i. steam boiler and no license is required to operate this boiler. Under these circumstances, you are allowed to adjust the safety valves.
a. true
b. false

29. The National Board requires a 'police' type safety valve be installed on a pressure vessel. This valve has
a. no external adjustments
b. a blow back setting only
c. 'pop' point setting only
d. 'b' and 'c'

30. When installing a boiler tube, the tighter you roll the tube the better the seal will be and the longer it will last. So, you should roll tube to maximum expansion capacity.
a. true
b. false

31. After rolling a firetube you can bead the end over with a
a. heavy lead mallet
b. heavy copper mallet
c. ball-pean hammer
d. none of the above

32. Chief Engineer instructs you to weld a boiler part that will be subjected to high pressure. With his signed written work order you should proceed to do the job and then later put the boiler on line.
a. true
b. false

33. A key-cap is a tapered plug sold by the boiler company for isolating a burst watertube. You may use a key-cap instead of a
a. soft pine plug
b. handhole cover
c. manhole cover
d. circulating nipple

34. ASME certified welder is required to repair steam and water drum cracks, but a leaking manhole cover can be welded without certification requirements.
a. true
b. false

35. You removed a jet reaction type safety valve on superheater. The only available replacement is one of the following. Which one will you use?
a. nozzle reaction
b. huddling chamber
c. 'police' type
d. none of the above

36. A gauge glass with tiny hairline scratches on the outside surface is okay to use, but not scratches on the inside surface exposed to water and steam.
a. true
b. false

37. You will install a blowdown line with extra heavy steel pipe.
a. true
b. false

38. Can you use a 300 WOG (Water Oil and Gas) valve on a 300 p.s.i. steam line or boiler?
a. yes
b. no

39. You can use a ________ valve in a confined location where handwheel cannot be operated properly.
a. OS and Y
b. gate
c. non-rising stem
d. globe
e. 'a' and 'b'

40. Where would you install the intake air filter and pipeline on a large air compressor in relation to the machines location?

a. outside building
b. inside building close to compressor
c. no air filter is required
d. adjacent to heat source to preheat air

41. When starting a package air compressor (for test run) with integral oil pump you must certainly first check

a. shaft rotation
b. air filter
c. aftercooler
d. safety valves

42. When installing a new blow down tank you must, before blowing down boilers

a. hydrostatic test the tank
b. ensure a water seal inside tank
c. test safety valve
d. bypass blowdown tank to sewer

43. Oil grooves in a sleeve bearing are to prevent shaft from squeezing oil from surfaces and creating metal-to-metal contact.

a. true
b. false

44. Where are split sleeve bearing oil grooves located in horizontal installation in most situations?

a. top bearing halves
b. bottom bearing halves
c. 'a' and 'b'
d. none are used

45. When cleaning equipment, (for example a motor driven centrifugal compressor) don't get motor damp unless the machine is running.

a. true
b. false

46. When installing a suction line to pumps from an underground tank. It is best not to

a. take suction from tank bottom
b. take suction at least 5" from tank bottom
c. install a check valve
d. install a suction line strainer

47. To tighten a 12-bolt flange, use _______sequence in the first four steps (answer in relation to clock dial).

a. 12 - 4 - 6 - 8
b. 12 - 6 - 3 - 9
c. 12 - 6 - 2 - 7
d. 12 - 1 - 2 - 3

48. Stagger the joint gaps 90 degrees when installing Lantern rings or packing into a stuffing box.

a. true
b. false

49. A valve for a 350 p.s.i. saturated steam line should have _______ stamped on the valve body.

a. 350 WOG
b. 400 WOG
c. 350-S
d. ASME 350 p.s.i.
e. National Board 350 p.s.i.
f. 'd' or 'e'

50. Testing a refrigeration system a halide blow torch will detect a leak as the Freon will change the flame color, but beware of toxic and deadly nerve gas being produced.

a. true
b. false

51. Natural gas pressure regulators have a vent hole in the upper diaphragm casing. It is best to vent this via piping outside of the building if this valve is located indoors.

a. true
b. false

52. Common packing end-gap cut(s)

a. scarf
b. square
c. step
d. all of the above

53. When a 10 ring stuffing box leaks, and packing gland nuts are bottomed to maximum tightness

a. loosen gland nuts, run machine, then tighten evenly
b. pull out 2 rings and install 2 new rings
c. lubricate packing with high temperature grease, perform 'a'
d. replace all 10 rings

54. Use crocus cloth when cleaning a finish ground joint, also soap stone can be used with oil, but has a tendency to scratch metal.

a. true
b. false

55. Chief Engineer instructs you to perform a repair or installation that may be contrary to the boiler pressure vessel code. You may be held responsible in case of accident if you perform the job. Therefore, you had better learn the boiler codes.

a. true
b. false

Chapter 6

Physics
500 HP

1. Exhaust steam

a. waste and not to be used
b. low quality
c. de-superheated
d. low pressure, quality good

2. Live steam

a. steam in motion
b. steam in steam line
c. taken directly from boiler
d. steam performing work

3. Corrosion in return lines is caused by

a. carbonic acid and water
b. carbon dioxide
c. oxygen
d. carbonic acid and 02

4. Water will gain its lubricating properties as temperature rises.

a. true
b. false

5. Adiabatic compression results if a gas is increasingly compressed with the gas temperature rising accordingly.
a. true
b. false

6. Deareating water can be accomplished by using
a. vacuum
b. pressure
c. trays and drip pans
d. all of the above

7. Steam is a
a. heated, compressed, visible vapor generated by a boiler
b. compressed heated vapor
c. heated vapor
d. invisible gas

8. Dynamic head
a. water in motion
b. water in a horizontal pipe-no pressure
c. water in a vertical pipe with pressure and no motion
d. refers only to low viscosity liquids

9. The following answer(s) that requires physical movement of a body to accomplish heat transfer
a. radiation
b. conduction
c. convection
d. a-c

10. Applying pressure to a block of ice will _____ the ice
a. freeze
b. melt
c. expand
d. compress

11. Combustion is uniting or combining
a. hydrogen, nitrogen and oxygen
b. carbon, nitrogen and oxygen
c. carbon and oxygen
d. nitrogen, carbon and oxygen

12. The 'heat of fusion' is the heat added or removed from a substance to change its state. What state is being changed?
a. solid to a liquid
b. liquid to a solid
c. liquid to a gas
d. gas to a liquid
e. 'a' and / or 'b'

13. Placing a razor blade gently on the surface of a body of water will actually float. What is responsible for this fact?
a. metal produces repelling electrons
b. metal has displaced its weight in the fluid
c. surface tension of the fluid
d. all of the above.

14. Low p.s.i. is read at ___________atmospheric pressure at sea level
a. 15.7 psi gage pressure
b. 114.7 absolute pressure in inches of water
c. 29.7 absolute pressure in inches of mercury
d. 15.7 psi reading on a compound gage

15. Oxygen in the atmosphere is (zero humidity 32°F)
a. at the low temperature vapor state
b. desuperheated
c. superheated
d. an unstable gas

16. A British thermal unit
a. 1 lb. of water lowered it's temperature 1° Farenheit
b. 1 lb. of water raised its temperature 1° Kelvin
c. 1 lb. of water evaporated to steam from 212° to 213° F
d. none of the above

17. Matter in motion contains
a. potential energy
b. kinetic energy
c. electrical energy
d. acceleration

***18.* Soft water in boilers**
a. is not acidic
b. has no minerals
c. prevents foaming and priming
d. can still cause water problems, deposits on tubes, etc.

19. Steam quality determined by
a. its temperature
b. its dryness
c. its chemical combinations
d. its ability to condense

20. A fluid is a
a. gas
b. liquid
c. 'a' or 'b'
d. semi-plastic state

21. Static head.
a. water in motion
b. water in a horizontal pipe with no pressure
c. water in a vertical pipe with pressure and no motion
d. refers only to high viscosity fluids

22. Natural draft is crested by
a. radiant heat
b. conduction
c. convection currents
d. conduction and radiant currents

23. Zero on a pressure gauge
a. 14.7 p.s.i.
b. atmospheric pressure
c. zero pressure
d. zero minus 14.7 p.s.i.

24. Flash point of fuel oil
a. when oil vaporizes
b. when oil burns steady
c. when oil fumes ignite
d. used to determine oil viscosity

25. Viscosity
a. low resistance to flow measurement
b. high resistance to flow measurement

c. medium resistance to flow measurement
d. resistance to flow measurement

26. Atom that loses or gains an atom is
a. split
b. ionized
c. fused
d. none of the above

27. One cubic foot of water will evaporate to a average volume of
a. one cubic foot
b. fifteen cubic feet
c. 1,728 cubic feet
d. One-tenth cubic foot

28. Water is lifted by suction.
a. true
b. false

29. One foot column of water, no matter the diameter of pipe, will have a pressure of .433 p.s.i..
a. true
b. false

30. Converging nozzles increase fluid pressure.
a. true
b. false

31. Diverging nozzles decrease fluid pressure and increase velocity.
a. true
b. false

32. Water temperature can be raised above 212°F in a steam vessel although vessel is vented to atmosphere.
a. true
b. false

33. Water hammer is also caused by hot and cold water making contact.
a. true
b. false

34. The highest temperature water will be in a vented atmospheric condition at sea level
a. 219° F
b. 212°F
c. 236°F
d. 120°F

35. Heat applied to a substance
a. molecules vibrate
b. molecules bonds tend to loosen
c. material will expand
d. all of the above

36. A solid stock of steel at ambient room temperature. Molecules are
a. frozen with no movement
b. still in motion
c. containing no heat energy
d. generating magnetic fields

37. Absolute zero.
a. - 460° F
b. all molecular movement ceases
c. no heat present
d. all of the above

38. A liquid will always
a. seek its own level
b. conform to any shape within container
c. have a higher temperature than a solid
d. all of the above

39. A gas.
a. molecular bonds no longer exist
b. can be compressed and confined in container
c. have a higher temperature than a liquid of same material
d. all of the above

40. Condensation of a gas occurs when molecules under pressure (such as in a steam boiler) are confined and in contact with its originating liquid, with no increase of heat applied.
a. true
b. false

41. Matter is made up of 92 basic elements.
a. true
b. false

42. Elements chemically joined are compounds.
a. true
b. false

43. Compound that can exist without breaking down into a element is a molecule.
a. true
b. false

44. Exhaust steam from most steam engines
a. waste and not to be used
b. low quality
c. desuperheated
d. low pressure, quality good

45. 'Live steam' is known to be
a. steam in motion
b. steam in any steam line
c. none of the above
d. steam releasing heat

46. Acidic corrosion in return lines is caused by
a. carbonic acid and water
b. carbon dioxide
c. oxygen
d. carbonic acid and 0^2

47. P.H. scale range
a. 0 - 7
b. 0 - 14
c. 0 - 10
d. 7 - 14

48. Water boils at sea level.
a. 212°F
b. 412°F
c. 390°F
d. 180°F

49. Degassing water can be accomplished by using
a. vacuum
b. pressure
c. trays and drip pans
d. all of the above

50. Desuperheated Steam is
a. Low temperature steam
b. compressed heated vapor
c. Steam with temperature higher than corresponding pressure
d. always an invisible gas

51. Dynamic head is a term to describe
a. a liquid in motion under pressure
b. flowing water in a horizontal pipe and under high pressure

c. high pressure water in a horizontal pipe with motion

52. PH is a measurement of

a. water density
b. oil and impurities in steam drum
c. acids and alkalines
d. steam quality
e. Percentage of Hydraulic pressure

53. Soft water.

a. water without chlorides
b. magnesium and calcium minerals converted to soft sodium
c. silica, iron, sulfate, and phosphorous removed
d. non-condensable gases removed
e. None of the above. Soft water allows boilers to generate more steam with less fuel usage.

54. Combustion is uniting or combining

a. hydrogen, nitrogen and oxygen
b. carbon, nitrogen and oxygen
c. carbon and oxygen
d. nitrogen, carbon and oxygen

55. Condensate corrosion is controlled by

a. ammonia
b. hydrazine
c. amines
d. alkalines
e. all of the above

56. Heat flows by

a. convection
b. conduction
c. radiation
d. all of the above

57. Low p.s.i..

a. Boiler under 400 square feet of heating surface operating at 16 p.s.i.
b. Pressure of Steam in Inches
c. Pressure under 15 p.s.i. gage pressure
d. 15 p.s.i. on a compound gage

58. When steam in its travels stops, it

a. creates water hammer
b. increases to a higher pressure
c. condenses
d. reverses its flow direction
e. produces a shock wave
f. 'a' and 'e'

59. British Thermal Unit

a. BTU = thermal measurement
b. BTU = thermal shock
c. 1 pound of water evaporated to steam from 212° to 213°F
d. measures heat gain but no losses
e. 'a' and 'd'

60. Turbine blades and shaft subjected to steam throttle opening =

a. potential energy
b. kinetic energy
c. electrical energy
d. acceleration

61. Soft water in firetube boiler
a. prevents caustic embrittlement
b. has no minerals to form scale deposits on tubes and shell
c. prohibits foaming and priming
d. prevents scale from forming

62. Steam quality =

a. high temperature steam
b. dry, low moisture steam
c. low acid steam
d. high resistance to condense

63. A fluid is also a ______ when molecular bonds non-exist.

a. gas
b. liquid
c. both 'a' and 'b'
d. plasma

64. A liquid is a

a. fluid
b. gas
c. both 'a' and 'b'
d. semi-plastic state

65. 'Static head' in a boilers blowdown line

a. water in motion
b. water in a horizontal pipe under no pressure
c. water in a vertical pipe under pressure and no motion
d. refers only to high pressurized water

66. Natural draft is created by

a. radiant heat
b. conduction
c. convection currents
d. conduction and radiant currents

67. Zero on a pressure gauge is always

a. 14.7 p.s.i.
b. atmospheric pressure
c. zero pressure
d. 14.7 minus gage p.s.i.

68. 'Flash point' of fuel oil

a. when oil vaporizes
b. when oil burns steady
c. when oil fumes ignite
d. used to determine oil viscosity

69. Viscosity.

a. low resistance to flow measurement
b. high resistance to flow measurement
c. medium resistance to flow measurement
d. resistance to flow measurement

70. Symbol for nitrogen.

a. N
b. $N0^2$
c. NO_3
d. Ni^2

71. Symbol for hydrogen.

a. H^2
b. H^2
c. H
d. HY

72. Symbol for carbon dioxide.

a. CO^2
b. CO^3
c. $C0_2$
d. CO

73. Symbol for methane the chief component of natural gas.

a. CH_4
b. CH
c. CH14
d. ME

74. Symbol for oxygen.

a. 02
b. 0_2
c. 0
d. 0^2

75. Symbol for carbon monoxide.

a. CM
b. CMO^2
c. CO
d. CO^3

76. Symbol for water.

a. H_20
b. HO^2
c. H^2O

77. Symbol for soft water.

a. H_20
b. $H0^2$
c. H^2O

78. Symbol for demineralized water.

a. H_20
b. H_2O_d
c. HO

79. One revolution of a circle.

a. 90°
b. 120°
c. 180°
d. 360°

80. The output of a polyphase electric generator is much higher than a single phase generator operating at same speed.

a. true
b. false

81. U-tube manometer is more sensitive than an incline manometer for measuring pressure changes

a. true
b. false

82. One BTU is equivalent to

a. 778 foot pounds
b. one horsepower
c. 33,OOO foot pounds
d. 'b' and 'c'

83. Calorie is a measure of heat applied or removed and in quantity can be related to BTU's.

a. true
b. false

84. A ship will float on water because of ______ theory.

a. ship's hull is lighter than water
b. air pockets in ship's metal
c. displacement
d. surface tension and 'a'

85. Helium balloon rises because

a. atmosphere exerts an up-lifting force
b. balloon is lighter than air
c. gravity has no effect on helium
d. displacement is taking place
e. 'a' and 'd'

86. A dielectric is a______ in a electrical device.

a. semiconductor
b. insulator
c. grounding wire
d. shunt circuit

87. An alternator will generate __________electricity.

a. AC power
b. DC power
c. 'a' and 'b'

88. Energy can neither be created nor destroyed, is the law of

a. Newton
b. Ohm
c. thermocouple
d. conservation

89. It is inappropriate to release fluorinated hydrocarbons into the atmosphere.

a. true
b. false

90. The symbol for nitrous oxide a gas routinely used in hospitals.

a. N_2O
b. NO
c. NO^2
d. O_2N

91. Symbol for ammonia.

a. NH^3
b. NH_3
c. OA
d. ammonia has no symbol

Chapter 7

Pumps 500 HP

1. To changeover pumps

a. stop one, start the other
b. start the other, trip off the existing
c. open vents, start existing
d. open vents, start pump

2. Cavitation can occur in turbulent hot water only.

a. true
b. false

3. To find 'static head' pressure of a deareator 50 feet above a boiler feed pump. Use which formula?

a. 50 divided by 2.44
b .2.44 divided by 244
c. multiply .433 times 50
d. multiply 2.44 times 50

4. The minimum height is________feet for a deareator to be placed above a boiler feedpump.

a. 5
b. 7
c. 9
d. 10

5. To prime any pump

a. open vent and increase speed of pump
b. open vent and backflow water through the pump
c. close vent and backflow
d. open vent, add water to suction line, close vent

6. A recirculating line on centrifugal pumps.

a. to boost performance
b. protect pump from overheating at low loads
c. regulate pressure only
d. used to prime pump under full-load

7. To start a feed pump with a leaky boiler check valve.

a. start pump quickly
b. start slowly with pump vent open

c. pump vent open, discharge valve closed
d. Pump vent closed, suction valve half-open

8. Capacity of a pump.
a. energy used
b. efficiency
c. gallons per minute
d. gallons per hour x energy consumed

9. Reciprocating steam pump valves have no
a. lap and lead
b. ports
c. cushion valves
d. slide valves

10. Each stage on a centrifugal pump will have a
a. recirculation line
b. vent
c. relief valve
d. all of the above

11. All pumps must be started
a. loaded
b. unloaded
c. quickly
d. 'b' and 'c'
e. none of the above

12. The most thermally efficient.
a. duplex piston steam pump
b. rotary pumps
c. simplex piston steam pump
d. inspirator
e. centrifugal pumps

13. A centrifugal pump will not pump to maximum rated specifications if
a. steam bound
b. air bound
c. running with pump recirculation valve open
c. running with automatic recirculating line valve open
d. discharge or suction line valve partly open
e. all of the above

14. To relieve a steam locked centrifugal turbine driven pump.
a. slow down pump, open vent, add cold water to deareator and / or pump casing
b. trip turbine, add heated water to receiver
c. add cool water from city or makeup supply to pump discharge line
d. all of the above is correct

15. To relieve any air locked pump with a check valve.
a. open pump vent and hose pump with cool water
b. open pump vent, fill DA tank or receiver with water, close vent
c. close discharge valve, open pump vent and fill receiver with water
d. speed up pump to increase vacuum to prime pump

16. To start a pump with a leaky check valve
a. slow down pump and speed up quickly
b. open recirculating line and throttle it slowly to close
c. slow down pump, open vent, close discharge, close pump casing vent
d. close discharge, open pump vent, start pump, close vent, open discharge valve when p.s.i. equalizes

17. 6 x 4 x 6 = pertaining to reciprocating pumps.

a. steam piston area x water piston area x number of strokes per minute
b. steam piston area x water piston area x length of stroke
c. steam piston area x length of stroke x water piston area
d. none of the above

18. An 'inspirator' is operated in theory like a
a. steam trap
b. reducing valve
c. centrifugal pump
d. injector or ejector
e. steam turbine

19. Centrifugal pump develops ______in discharge line.
a. pressure
b. head
c. hydraulic force
d. potential energy

20. Rotary pumps are used mostly for
a. high viscosity liquids
b. low viscosity fluids
c. high viscosity fluids
d. ultra low viscosity fluids that are inflammable

21. What is found on fuel oil pump suction line?
a. simplex strainer
b. duplex strainer
c. triplex strainer
d. relief valve

22. How can you tell the difference between a suction and discharge line?
a. discharge line is larger
b. suction line has a gate valve
c. relief valve on discharge
d. both lines are identical

23. Discharge line on a pump has a
a. safety valve
b. relief valve
c. duel strainer (s)
d. quick opening straight away valve

24. Suction line on a pump has a
a. safety valve
b. vacuum breaker
c. relief valve
d. strainer

25. Once in operation, an automatic injector will start by itself if vacuum is momentarily broken in suction line.
a. true
b. false

26. Can you run a centrifugal pump without water?

a. yes, if casing vent is open
b. no
c. yes, but only when testing a new pump

27. Pump that has no moving parts when operating
a. injector
b. ejector
c. jet pump
d. inspirator
e. all of the above

28. How many steam ports in a duplex reciprocating pump with 'D' type valves?
a. 4
b. 5
c. 10

29. How many valves in a centrifugal duplex single stage slide valve pump?
a. 8 or more
b. 10 or more
c. no valves in this pump
d. no such pump exists

30. Inspirators will pump hot water.
a. true
b. false

31. To changeover pumps
a. stop one, start the other
b. start the other, trip off the existing
c. open vents and start existing
d. open vents and start pump

32. A multistage feed pump has
a. one impeller
b. two or more impellers
c. two shafts
d. one impeller and two or more stages

33. In a two-stage pump. Suction or first stage is
a. smaller
b. same size
c. larger
d. none of the above

34. Check valve on a boiler feed pump is installed
a. on suction line
b. in recirculating system
c. discharge line before discharge valve
d. on the discharge line after the pump's discharge valve

35. Reciprocating pump discharge line starts pounding and shaking violently. You should check the
a. piston rings or packing
b. suction valves
c. suction strainer
d. air chamber
e. all of the above

36. The exhaust steam for a turbine driven pump will contain a trace of oil, but not as much as a duplex steam pump.
a. yes
b. no

37. How can you tell a suction valve is leaking on a reciprocating duplex steam pump with no footvalve placed in suction line?
a. pump would stop
b. piston shafts would short stroke
c. one piston stroke will be accelerated
d. you couldn't tell until an internal inspection of the valve

38. If you see a steam line entering the water-end of a pump with a trap and condensate line leaving, this pump would be a ______pump .
a. condensate
c. vacuum
c. steam jacketed
d. injector

39. Cavitation will result from overspeeding a pump. The suction draft is so strong that cavities (holes) actually appear in the water which can damage the pump.
a. true
b. false

40. If pump will build up pressure, is noisy, and will not pump.
a. pump is air bound
b. pump is steam bound
c. pump has a bent shaft
d. impeller frozen to shaft

41. You will most likely find a ____________ on the discharge line on a pump.
a. compound pressure gage
b. flow meter or flow chart sensor
c. safety valve
d. all of the above

42. A pump found on a steam condenser to maintain vacuum.
a. jet
b. vacuum (rotary)
c. reciprocating
d. any of the above can be used

43. Boiler feed pump is running fine. A small turbine is used to supply electricity for the power plant and exhaust steam is used to supply the deareator. No reducing valves are found in this plant. You started to shed electrical load on the turbine a bit too quickly and noticed boiler feed pump began cavitating. What caused this?
a. 'total' head pressure drop to the pump
b. DA temperature increased
c. DA pressure increased
d. boiler feed regulator still demanding previous load
e. DA pressure decreased causing 'a' and 'd' to result

44. A pressure regulator has a handwheel located above the diaphragm. This handwheel can be used to adjust the pressure on the downstream side even though a constant speed centrifugal pump is supplying fluid on the upstream side to the regulator.
a. true
b. false

45. Water will boil at a lower temperature in a vacuum. This is why a pump can become steam bound.
a. false
b. true

46. You have a pump that constantly becomes steam bound. You could _______ to help remedy the problem.

a. inject cold water in suction line

b. add cold make-up to receiver

c. reduce steam pressure in DA

d. all of the above

47. Centrifugal pump suddenly loses all pressure.

a. check valve broke, blocking discharge line

b. casing vent dripping water

c. automatic recirculation valve closes

d. Impeller slipping on shaft

48. How would you prime a reciprocating pump that is drafting fluid from a full supply tank, but the footvalve on the suction line is leaking due to corroded valve seats?

a. fill suction line then start pump

b. place a vacuum pump on suction line then start pump

c. open vent and run pump

d. 'a' or 'b'

a. none of the above. Replace footvalve

49. The exhaust steam from a duplex steam pump will contain oil.

a. false

b. true

50. A 'rupture disk' can be found on a pump's discharge system, but should be piped to release fluid to a safe point of discharge as a relief valve also should.

a. true

b. false

51. A condensate pump taking suction from a vented receiver. The receiver should be at least ________feet above the pump.

a. 1 1/2

b. 3

c. 6

d. 7

52. A boiler feed pump requires a 10 p.s.i. static head at the suction. How high would you install the deareator above the pump?

a. 5-feet

b. 8-feet

c. 20-feet

d. 50-feet

53. For a given rated flow and pressure, centrifugal pumps are smaller in size than reciprocating or rotary pumps.

a. false

b. true

54. When operating a pump passing water into a closed heater it is most important to routine check the

a. trap drain on heater

b. steam pressure

c. water outlet temperature

d. all of the above

55. Large multistage centrifugal pumps are also referred to as _______type pumps.

a. barrel

b. medium speed

c. bucket wheel

d. rotary

56. The main purpose of a suction line strainer on a centrifugal pump is to prevent

a. uncontrolled flow

b. bearing failure

c. erosion

d. all of the above

57. If the wearing rings are worn, the discharge will short circuit to the suction side in a centrifugal pump.

a. true

b. false

58. You need to install a positive displacement pump. Which one would you purchase if flow is needed to be constant?

a. centrifugal

b. reciprocating

c. power pump

d. rotary

e. diaphragm

59. Which of the following pumps will have a renewable liner?

a. injector

b. reciprocating simplex

c. reciprocating duplex

d. return trap

e. 'b' and 'c'

f. none of the above

60. An 'eccentric reducer' pipe fitting to a pump will be located

a. on discharge line

b. on suction line

c. on discharge chamber

d. on drive shaft coupling

61. Water can be compressed

a. true

b. false

62. The stages in a centrifugal pump will be ______in size from suction to discharge.

a. larger

b. smaller

c. equal

63. Staybolts on a large centrifugal multistage pump would be to

a. detect a leak in a stage or along the shaft

b. tie and support stage sections to a single package

c. align shaft axial clearance

d. all of the above

64. A device used to help take load off of thrust bearing in a large multistage centrifugal pump is a _____. If this were not installed, a very large thrust bearing would be needed to compensate.

a. balancing disk

b. balancing drum

c. dummy piston

d. any of the above

65. To install a pump liner, you will have to use carbon dioxide to shrink-fit into case. Therefore, frozen carbon dioxide in a solid state is referred to as

a. carbonic silica block ice

b. dry ice

c. CO_2

d. all of the above

66. Water, including hydraulic oils, can be slightly compressed if gasses are present in solution.

a. true

b. false

Chapter 8

Electricity 500 HP

1. Electricity flows in a circuit because of

a. electrons unevenly distributed
b. gravity forces electron to ground
c. negative electrons and protons always attract
d. 'a' and 'b'

2. One coulomb flowing in a conductor for one second is a

a. watt
b. volt
c. ampere
d. micro hertz

3. Resistance to electron flow is measured by

a. ohm meter
b. volt meter
c. megohm meter
d. 'a' and 'c'

4. The larger the conductor the less resistance and heat.

a. true
b. false

5. Resistance increases as temperature increases.

a. true
b. false

6. A good conductor of electricity is

a. air
b. glass
c. brass
d. porcelain

7. Opposition to the flow of electrons in a substance is a

a. conductor
b. resistance
c. short circuit
d. ground fault

8. Positive charged particle of electricity.

a. neutrons
b. proton
c. nucleus
d. electron

9. Negative charged particle of electricity.

a. electron
b. neutron
c. proton
d. nucleus

10. A coulomb is an electrical charge transferred in one second by a current of one ampere; i.e., it is 1 ampere-second.

a. true
b. false

11. American wire gage #12 is smaller than AWG 10.

a. true
b. false

12. Circular mil = cross sectional area of a .00111 wire's diameter.

a. true
b. false

13. Voltage

a. potential flow
b. potential difference
c. reluctance
d. all of the above

14. Potential difference creates a force called

a. watts
b. eddy currents
c. hysteria
d. electromotive

15. To check insulation resistance on a motor use a

a. wattmeter
b. anmeter
c. megger
d. growler

16. Sparking and possible flashover on a dynamo's commutator is caused by

a. brushes too tight
b. brushes too loose
c. dynamo speed 10% below normal
d. all of the above

Chapter 9

Turbines 500 HP

1. A steam turbine.

a. mechanical energy to electrical energy
b. steam to impeller blades
c. kinetic to mechanical energy
d. rotating potential energy to blades

2. A water turbine.

a. high density fluid energy transfer
b. kinetic to potential energy transfer
c. built just like a steam turbine
d. high density potential energy transfer

3. Oil in boiler-running a turbine

a. labyrinth seals broke
b. oil cooling bearing jacket cracked
c. bent rotor
d. none of the above

4. A steam engine or turbine can safely and reliably run on

a. vapor
b. any hydraulic fluid
c. air
d. hydrogen gas

5. Before starting a turbine driven feed pump

a. check oil
b. check exhaust valve
c. check ceasing drains
d. all of the above

6. Impulse turbine.

a. steam is deflected
b. steam strikes at right angle
c. steam enters at 45 angle
d. steam strikes at 60 angle

7. Reaction turbine.

a. steam is deflected
b. steam strikes at right angle
c. steam loses velocity when leaving diaphragm
d. steam reacts with working force in diaphragm rotation

8. Regenerative turbine.

a. impulse type
b. opposed flow type
c. reaction type
d. single stage with standard shrouded blades

9. Wetsteam in an operating turbine-

a. makes noise
b. lowers speed
c. increases speed
d. lowers oil temp

10. An atmospheric relief valve is found on turbine
a. second stage
b. exhaust line
c. main steam line
d. drains

11. A steam turbine is a
a. heat engine
b. pressure rotary engine
c. reducing engine
d. rotary compressor engine

12. Main steam line to a prime mover must be
a. inclined toward boiler
b. inclined toward prime mover
c. level all the way
d. in a continuous loop in all situations

13. A noncondensing t turbine exhaust steam contains
a. engine oil
b. moisture
c. cylinder oil
d. turbine oil and moisture
e. none of the above

14. Turbine labyrinth seals are lubricated by
a. water
b. steam
c. oil
d. all of the above
e. none of the above

15. Superheated steam in a turbine increases efficiency by
a. blades absorbing more heat into rotor
b. increasing speed
c. reducing back pressure on first and middle stages
d. reducing friction and excess condensation

16. Shutting down a turbine driven feed pump, first close
a. discharge valve
b. steam exhaust valve
c. steam valve
d. casing drains and steam valve

17. The overspeed trip
a. automatic and adjustable
b. manual, automatic, and fully adjustable
c. manual and can be adjusted only with turbine at low load
d. cannot be adjusted, and is not manually operated

***18.* Turbine throttle valves.**
a. gate
b. balanced double seat
c. double seat gate
d. balanced double seat gate with split wedge

19. Expansion joints on steam lines to and from turbine are to minimize stress from expansion and contraction.
a. true
b. false

20. How can you verify a turbine tachometer is accurate when the turbogenerator is generating electricity?
a. check voltage
b. verify amperes
c. check hertz
d. all of the above

21. A steam turbine.
a. mechanical energy to electrical energy
b. steam to impeller blades
c. kinetic to mechanical energy
d. rotating potential energy to blades

22. A hydraulic water turbine.
a. high density fluid energy transfer
b. kinetic to potential energy transfer
c. built and designed exactly like a steam turbine
d. high density potential energy transfer

23. Oil in boiler running a turbine. What is wrong?
a. labyrinth seals broke
b. oil cooling bearing jacket cracked
c. governor malfunction
d. none of the above

24. A steam engine or turbine can run on
a. vapor
b. any hydraulic fluid
c. air
d. hydrogen gas

25. Before starting a turbine driven feed pump.
a. check oil
b. check exhaust
c. check casing drains
d. all of the above

26. Impulse turbine.
a. steam is deflected
b. steam strikes at right angle
c. steam enters at 45°angle
d. steam strikes at 600° angle

27. Reaction turbine.
a. steam is deflected
b. steam strikes at right angle
c. steam loses velocity when leaving diaphragm
d. steam reacts with working force in diaphragm rotation

28. Regenerative turbine.
a. impulse type
b. opposed flow type
c. reaction type
d. single stage with standard shrouded blades

29. Wet steam in a operating turbine
a. makes noise
b. lowers speed
c. increases speed
d. lowers oil temp

30. Atmospheric relief valve is found on turbine is found on
a. second stage
b. exhaust line
c. main steam line
d. casing drains

31. A steam turbine is a
a. heat engine
b. pressure rotor engine
c. reducing engine
d. rotary compressor engine

32. Main steam line to a prime mover must be
a. inclined toward boiler
b. inclined toward prime mover
c. level all the way
d. in a continuous loop in all situations

33. Noncondensing turbine exhaust steam contains
a. engine oil
b. moisture
c. cylinder oil
d. turbine oil and moisture
e. none of the above

34. Turbine labyrinth seals are lubricated by
a. carbon rings
b. steam
c. grease or oil
d. all of the above
e. none of the above

35. Superheated steam in a turbine increases efficiency by
a. blades absorbing more heat into rotor
b. increasing speed
c. reducing back pressure on first and middle stages
d. reducing friction and excess condensation

36. Shutting down a turbine driven feed pump, first close
a. discharge valve
b. steam exhaust valve
c. steam valve
d. casing drains and steam valve

37. The overspeed trip.
a. automatic and adjustable
b. manual, automatic and fully adjustable
c. manual and can be adjusted only with turbine at low load
d. cannot be adjusted and is not manually operated

38. Turbine throttle valves.
a. gate
b. balanced double seat
c. double seat gate
d. balanced double seat gate with split wedge

Chapter 10

Auxiliaries 500 HP

1. Backflow preventer protects.
a. drinking water
b. boiler
c. feed pumps
d. turbine

2. Steam temperature alarm is located on
a. exhaust steam header
b. nonreturn valve
c. steam drum
d. superheater outlet

3. A deareator will also remove nitrogen gas.
a. true
b. false

4. Orsat device measures CO, 0_2 and CO_2 in the
a. last pass
b. flue
c. economizers and first row of tubes
d. combustion chamber

5. Air reheaters are located before the economizer.
a. true
b. false

6. Cooling tower water is treated to prevent
a. foaming
b. priming
c. excessive blow down of boilers
d. algae growth and corrosion
e. 'c' and 'd'

7. Cooling tower water PH should be between
a. 3 to 5.6
b. 4.7 to 5.0
c. 6.8 to 7.4
d. 9 to 11

8. What does a demineralizer do?
a. reduce calcium carbonate to a semi-fluid state
b. change CO_2 into 0_2 removing Calcium Carbonate
c. displace 0_2
d. remove minerals

9. The cation unit in a demineralizer is regenerated by
a. sulfuric acid
b. sodium nitrate
c. ammonia
d. no need to regenerate

10. The anion unit in a demineralizer is regenerated by
a. ammonia
b. soda ash
c. sulfuric acid
d. no need to regenerate

11. High water in a deareator.
a. reduces efficiency and temperature
b. corrodes the inside surfaces
c. Activates low-level alarm

12. To solve feedwater hammer and shock to piping, install
a. steam traps
b. air chamber or expansion tank
c. lower PH with acid
d. all of the above

13. Most oil and or steam separators function with
a. filters
b. baffles
c. centrifugal force
d. all of the above

14. If you were piping up a feed line, what would you install on it?
a. check valve
b. relief valve
c. pressure and or temperature gage
d. city water line
e. bypass around regulators
f. all of the above
g. only 'a' 'c' 'd' and a safety valve

15. Why aren't water columns insulated?

a. so water will be lower in glass due to density
b. will be problematic when glass breaks or when trycocks are opened
c. to prevent steaming in column and evaporation
d. so column won't rust

16. Condenser vacuum is measured in
a. inches of water
b. absolute pressure
c. inches of mercury
d. square inches of water
e. 'c' and 'b'

17. If you change steam flow direction through a reducing valve, it will become an increasing valve creating high pressure from low pressure steam.
a. true
b. false
c. true, depending if valve operates with a pilot and sensor

18. To desuperheat high pressure steam
a. inject low p.s.i. steam which is cooler
b. inject 0_2
c. inject demineralized water
d. reduce steam flow, increase firing rate in boiler

19. An open heaters accessories.
a. low water float makeup
b. overflow
c. relief valve or back pressure valve
d. alarm system
e. 'a' 'c' and 'd'
f. all of the above

20. A inverted bucket trap can get air or steam bound?
a. true
b. false

21. A bucket trap.
a. hinged upright bucket with a discharge tube and valve stem
b. inverted bucket on a hinge with valve
c. a moving flat disk
d. none of the above

22. The best steam trap to use on heat exchanger
a. thermodynamic
b. bucket
c. inverted bucket
d. float

23. If you had a closed feedwater heater, feed pump running at full capacity and boiler losing water.
a. steam trap on heater won't reseal and steam pouring out continuously
b. steam pressure too high in heater
c. back pressure valve will not close
d. broken coil or tube

24. Air injectors are used on
a. condensers
b. boiler steam drum
c. condensate return line to deareator
d. none of the above

25. Orsat samples
a. $C0_2$
b. 0_2
c. carbon monoxide
d. all of the above

26. Name a steam trap that can feed water to a boiler.
a. float
b. inverted bucket
c. return trap
d. thermodynamic

27. High water in an open feedwater heater is caused by
a. makeup control valve leaking
b. Makeup water valve open
c. boiler load reduction
d. all of the above

28. Low water in an open feedwater heater
a. boiler load increase
b. makeup valve closed
c. condensate pump failure
d. all of the above

29. High water in a closed feedwater heater.
a. makeup valve open
b. boiler load decrease
c. condensate pump running continuously
d. no water level in this heater

30. If a open heater dumps overflow water and you were standing near the drain discharge, would there be a warning first?
a. yes, the alarm will always activate to warn you
b. no, you could be scalded

31. All alarms and boiler accessories such as low water cut-outs are so reliable, as to be trusted upon to operate
a. true
b. false

32. What to do if cooling tower water foaming?
a. increase blow downs
b. use a foam depressant chemical
c. lower chemical concentrations
d. all of the above

33. Deareator water and steam temperature should be close to 3 degrees difference or better.
a. true
b. false

34. If water temperature in DA was dropping.
a. air accumulation in steam space
b. supply steam failing
c. cold makeup feeding at high capacity
d. all of the above

35. You have reason to believe DA tank has a air blanket condition. What would you do?
a. increase steam
b. decrease steam

c. open vent valve
d. add makeup water

36. Water sprayed into the upper steam space of a deareator is called the

a. vent condensing section
b. preheating section
c. primary separation section
d. all of the above

37. What would make a deareator safety valve open?

a. run away turbine
b. reducing station not throttling
c. low water
d. 'a' and 'b'
a. all of the above

***38.* What will cause a vacuum in a deareator?**

a. leaking gage glass
b. condensate flow to heater below normal
c. condensate flow to heater above normal
d. steam flow to heater too low
e. 'c' and 'd'
f. all of the above

39. The vent valve on a dearator should be

a. throttled
b. issuing some steam
c. closed
d. 'a' and 'b'

40. To locate air leak in a condenser, use

a. candle flame
b. smoke
c. leak detector fluid, it will form bubbles
d. 'a' or 'b'
e. vacuum sensor probe

41. A shell and tube heat exchanger can be applied to

a. condense gas
b. heat a fluid
c. cool a fluid
d. evaporate a liquid
e. all of the above

42. Shell and tube heat exchangers are classified by

a. counterflow
b. parallelflow
c. crossflow
d. all of the above

43. A reheater can heat steam by utilizing a

a. pass into a boiler furnace
b. steam to steam heat exchanger
c. separate fired heat exchanger
d. all of the above

44. Most steam flow meters will have____ indicators.

a. flue gas
b. steam flow
c. airflow
d. all of the above

45. Shut down a condenser.

a. secure gas medium
b. secure fluid medium stop valve
c. secure fluid medium pumps
d. secure condensate pump
e. 'a' 'c' 'b' and 'd'
f. 'a' 'b' 'c' and 'd'

46. What should be installed on the downstream side of a reducing valve?

a. strainer
b. check valve
c. relief valve
d. all of the above

47. A condenser hot-well acts as a receiver for a pump to obtain suction and indicate condensate level.

a. true
b. false

48. If water kept rising in a condenser hot-well.

a. loss of vacuum
b. air ejector failed
c. leaking cooling water tube
d. feed line to boiler broke

49. ASME certified welders should not weld boiler metal if water or steam is present.

a. true
b. false

50. A flexible staybolt has a telltale hole if used on a vertical firetube boiler.

a. true
b. false

51. A steam engine has no overspeed trip due to the fact an engine has a cut-off governor to prevent this, but a belt driven throttling governor will have a trip device if belt falls off pulleys or breaks. However, neither governor on engines senses speed to activate a trip.

a. true
b. false

52. The governor on a steam reciprocating pump throttles the steam to the pump to control speed.

a. true
b. false

53. Drips are another term for drains on steam engine or pump.

a. true
b. false

54. In a jet type condenser, steam and water mix to exchange heat.

a. true
b. false

55. You are operating a impulse back-pressure turbine and load has been increased to a permanent level, but auxiliary exhaust line condenser is operating continuously since. What is the best method to insure high quality feedwater and maintain a high heat rate of plant efficiency?

a. install a demineralizer
b. install a vacuum deareator
c. install an evaporator
d. install an ion-exchange water softener

56. Where will you find a deareating hot-well?

a. condenser
b. suction line to feed pump
c. demineralizer
d. condensate polishing units

57. You have a plant with a deareator and closed feedwater heater. You're installing an injector for an emergency backup in case of a pump outage which are electrically driven centrifugal pumps, and another that is a high capacity steam driven pump. Where is the best location to connect the suction line from the injector?

a. to main makeup water line
b. injector can't be installed
c. to steam pump discharge line.
d. to steam pump suction line.

58. A shaft governor is used on

a. steam engines
b. steam pumps
c. turbines
d. diesel engines

59. Chart used to measure smokestack smoke density.

a. altimeter
b. manometer
c. Ringlemann
d. none of the above

60. A desuperheater on a reheat line to a turbine is mainly installed for emergency cooling, not for steam temperature control.

a. true
b. false

Chapter 11

Boilers Unlimited

1. Taking a flue gas analysis, low CO_2 means

a. perfect combustion
b. excess air
c. nitrogen being burned
d. fuel oil temperature too hot
e. flue temperature too hot

2. Is excess air desirable in small percentages during combustion of fuels?

a. yes
b. no

3. A furnace explosion can be caused by

a. pilot out of align or blown out
b. low volatile fuel at low load conditions
c. draft fan failure or flame detector malfunction
d. any of the above

4. A damper regulator controls

a. circulation in boiler
b. flue gas
c. water level
d. steam flow

5. An important feature a superheater safety valve has.

a. small size
b. no lifting lever
c. exposed spring
d. all of the above

6. Boilers over 500 square feet heating surface have

a. one safety valve
b. two safety valves
c. three safety valves

7. A superheater safety valve with no stop valves intervening between boiler and drum safety valve, can be included in calculating total steam relieving capacity of boiler.

a. true
b. false

8. A reheater safety valve with no stop valves intervening between boiler and drum safety valve, can be included in calculating total relieving capacity of the boiler.

a. true
b. false

9. A safety valve can be installed in a horizontal position.

a. true
b. false

10. You have reason to believe the water connection to the glass is obstructed and there is 1/2 glass of water showing. What do you do?

a. try gage cocks
b. blow down boiler to see if water level drops
c. raise steam pressure to free blockage
d. all of the above

11. Why are bent tubes used in watertube boilers?

a. so they won't crack
b. tubes enter drum at right angles
c. so scale can be trapped in tube and not enter steam drum
d. reduces tube length and costs

12. What is the predominate force to induce water circulation in steam boilers?

a. gravity
b. density differences
c. heat
d. steam pressure

13. Should you blow down water wall headers when boiler is on the line?

a. yes
b. no

14. What is a reheater on a water tube boiler?

a. economizer
b. air preheater
c. superheater
d. fuel oil heater

15. What is used to join the sections in a cast iron sectional boiler?

a. staybolts
b. welds
c. push nipples
d. rivets
e. 'a' and packing material

16. Since high pressure steam boilers under supervision don't require low water fuel cut-outs Do cast iron low pressure boilers require the

a. yes
b. no

17. A new boiler is rated maximum of 100 p.s.i. with a safety factor of 4. At what pressure will boiler most likely explode?

a. 350 p.s.i. to 390 p.s.i.
b. 375 p.s.i. to 400 p.s.i.
c. 400 p.s.i. to 500 p.s.i.
d. 204 p.s.i. to 350 p.s.i.

18. Boiler is 5-years in service running at 230 p.s.i.. MAWP is 250 p.s.i.. Can you run this boiler at anytime up to 250 p.s.i.?

a. yes
b. no

19. A riveted butt strap joint on a steam and water drum. Caulking is performed on the _____surface of the drum.

a. inside
b. outside
c. 'a' and 'b'
d. no caulking required

20. Do all staybolts require tell-tale holes?

a. yes
b. no

21. Hard water can cause foaming in a boiler.

a. true
b. false

22. Unusually high feed water pressure is necessary to force water in a boiler.

a. scale accumulation in feed line
b. valve broken loose in feed line
c. feed regulator partially open
d. Any of the above could cause this

23. When laying-up a boiler in hot summer weather.

a. store wet or dry method
b. clean soot from furnace
c. secure fuel and blow-off lines
d. all of the above

24. Untreated feedwater causes

a. scale
b. lost efficiency
c. corrosion
d. all of the above

25. Scale or solids in boiler water can cause tube erosion.

a. true
b. false

26. Most reliable test for a low water cutout.

a. blowdown sediment chamber
b. blowdown boiler with fires out
c. close feed valve, observe water level allowing cutout to trip fuel valves
d. disassemble cutout and place in bucket of water and test with electrical volt / continuity meter

27. If a staybolt blows steam, can it be welded to stop leak?

a. yes
b. no

28. Continuous blowdown controls

a. total dissolved solids
b. scale in mud drum
c. steam quality
d. 'b' and 'c'

29. An indicator that automatically measures smoke density.

a. calorie meter
b. orsat
c. photo cell light beam or electronic sensor
d. Ringleman chart

30. Excess air in large amounts

a. increases efficiency
b. wastes fuel
c. dirties the tubes
d. none of the above

31. Where would you install a flue gas thermometer?

a. just beyond the forced draft fan
b. where flue connects to stack after the flue damper
c. just before the flue damper
d. five feet from top of stack

32. What would you do if superheater safety valve did not open before drum safety?

a. shut down boiler fires only
b. shut down turbine or close stop valves and perform 'a'
c. lower firing rate, maintain water level
d. perform 'a' and keep turbine running

33. The top cover-gasket on a water column is leaking a small amount of steam.

a. tighten cover bolts in a cross pattern
b. tighten cover bolts in a circular pattern
c. relieve the pressure and perform 'a'
d. relieve the pressure and perform 'b'

34. You're working first day in a plant and all equipment is operating fine. The steam is used only on the industrial side. All of a sudden you instantly see the boiler water level rise to the top of the gage glass. What caused this?

a. foaming
b. large steam flow demand
c. feed regulator stuck open
d. burners tripped out

35. Your working first day in a plant and all equipment is working fine. The steam is used only on the industrial side. Suddenly you notice boiler water level rise to the top of the gage glass, then water instantly drops to 1-gage. What caused this?

a. foaming
b. low steam flow demand
c. feed regulator stuck open
d. all of the above

36. Boiler is steaming at 50% rating, feed supply fails. The burners will instantly (automatic combustion controls)

a. go to high fire
b. go to low fire
c. trip out
d. activate low water alarm

37. If safety valves popped open, what will the water in boiler do?

a. drop in gage glass
b. rise in gage glass
c. level stays the same

d. water will boil in gage glass

38. You test a safety valve by hand and it won't reseat.

a. leave valve alone
b. keep lifting handle, blowing valve
c. gag the valve
d. tap valve with rubber hammer

39. You blowdown a boiler. Both valves won't seat and you're rapidly losing water in boiler.

a. kill fire
b. open bypass to feed regulator
c. close stop valves
d. blowdown column and glass and try gage cocks
e. A and B

40. If you started a centrifugal pump and no water was entering the boiler, how could you verify feedwater flow other than watching the gage glass?

a. feed line discharge pressure
b. pump suction pressure
c. feedwater temperature
d. any of the above will verify

41. When inserting boiler on line the pressure was equalized. The nonreturn valve disk appears to be stuck closed.

a. raise pressure slowly, it will open
b. close down on handwheel and open up repeatedly.
c. take boiler off line, close outside stop and drain valve. Perform 'b'
d. close outside stop slowly and open drain valve

42. With full automatic flame safeguard system, is it possible to light burner off with a hand torch?

a. yes
b. no

43. Hot slag on firebrick is caused by

a. water in oil
b. burner out of align
c. oil too thin (low viscosity)
d. too little fuel, and too much air in furnace

44. Hot slag on firebrick could likely

a. fool flame detector and cause a furnace explosion
b. cause brickwork to collapse
c. can do no harm
d. overheat boiler furnace

45. Running a vertical boiler at full burner capacity with 1-gage of water, could

a. break staybolts
b. destroy upper tube sheet and tubes
c. melt fusible plug
d. produce more steam than safety valve's rated capacity

46. Two safety valves on a boiler. One is small the other is large. Which one should open first?

a. small valve
b. large valve
c. both at same time
d. both valves must be same size

47. A boiler is off line on wet standby. You open the drain to the gage glass and you see bubbles. What is wrong?

a. pressure in boiler drum
b. vacuum in boiler drum
c. main stop valves leaking
d. feed valve is leaking

48. You have 2-gauges of water. You blow the column and glass and see no water in glass.

a. extinguish burners
b. try gage cocks
c. blow column and glass again
d. all of the above in this order

49. You blow the low water cut- out and discover it doesn't work on the only boiler available.

a. take boiler off line immediately, call chief engineer
b. run the boiler under close surveillance
c. blow the column and glass
d. blowdown the boiler

50. On a riveted longitudinal seam on steam and water drum, the inside strap on butt strap joint is

a. larger
b. smaller
c. same as outside strap

51. Pitch of riveted joints refer to

a. angle of rivets when entering the sheets
b. center-to-center measurements of rivets
c. pitch of rivets as they are expanded into the sheets
d. two or more rivets diverging from a common point.

52. If you extinguish the burner on high capacity, the water in boiler

a. rises slowly
b. rises fast
c. lowers slowly
d. lowers fast

53. If you raise fires to maximum burner capacity, the water in boiler

a. rises slowly
b. rises fast
c. falls slowly
d. falls fast

54. Which boiler is dangerous with oil in it?

a. watertube
b. firetube
c. both are equally dangerous

55. When should you close the main stop valves when taking a boiler off the line?

a. anytime after killing the fire
b. when nonreturn valve cycles open and closed less than once per minute

c. when boiler acts as a condenser
d. when water in boiler drops in gage glass

56. When starting a boiler with superheater.
a. close superheater drains
b. open superheater drains
c. inject cold makeup water to cool tubes
d. lock open superheater safety valve

57. Operating a firetube boiler and a bag formed (manual stop valves installed).
a. shut down burners and close stops
b. shut down burners and lift safety's by hand
c. blow down boiler, then shut down burners
d. shut down burners only

58. Running a firetube and watertube boiler. Bag forms in firetube (manual stop valves installed).
a. shut down firetube boiler and isolate by closing stops
b. shut down watertube boiler and close stops
c. shut down burners in both boilers only
d. shut down firetube burners only

59. Operating cross-drum and longitudinal watertube boilers. Low water in longitudinal.
a. open bypass wide-open on longitudinal
b. crack open bypass on longitudinal and lower firing rate
c. lower firing rate on cross-drum
d. crack open bypass on longitudinal. Lower firing rates on both boilers

60. Introducing hard water in any boiler.
a. calcium carbonates will settle and cause overheating damage
b. hard water is okay if its been filtered
c. will pass 'hard steam' throughout system
d. none of the above

61. Three boilers on line have manual stop valves and tube blows with engine or turbine running. What would you do? (plant has only 1-feed pump, the other is broke)
a. kill fire on # 1. Open feed bypass
b. kill fires on all three boilers. Stop turbine and open feed bypass
c. kill fire on all boilers. Leave turbine running. Open feed bypass, throttle down feedwater to normal boilers.
d. kill fire on #1. Leave turbine running. Open feed bypass.
e. kill fire on all boilers. Blowdown all boilers to relieve pressure. Open feed bypass and stop turbine

62. When inserting boiler on line with manual stops, you hear a rushing sound.
a. open stop valve a bit more, a little at a time
b. open valve quickly
c. close valve
d. this is normal to hear. Perform 'a'

63. What caused the problem in question #62?
a. pressure not equalized
b. high water
c. not a problem
d. feedwater too hot

64. What caused the problem in question #63?
a. worn spindle on stop valve
b. defective steam gauge
c. drain valve between stops cracked open only one turn
d. fire too high in one of the boilers
e. not a problem

65. Firing on oil. Electric circuits okay on automatic burner controls. Flame fails.
a. atomization pressure too high
b. atomization pressure too low
c. dirty oil pump strainers or cold oil
d. all of the above

66. A few fire cracks in riveted joints.
a. are okay if cracks don't enter rivet hole area
b. very dangerous condition
c. cracks caused by low water
d. cracks caused by high alkaline water

67. If gage glass broke and there is no replacement.
a. shut down boiler
b. run boiler using gage cocks
c. 'a' and 'd'
d. call Chief Engineer

68. For best steam 'quality'
a. carry a lower water level
b. carry a higher water level
c. add no chemicals to boiler water
d. raise total dissolved solids above recommendations

69. Oil fired burner won't start.
a. flame detector dirty or cold oil
b. dirty burner tip or no vacuum in oil pump suction line
c. steam atomizing pressure out of range or wet steam
d. all of the above

70. Two boilers on line. Feed pump working okay. #1 boiler has 1-gage of water. #2 has 3-gages. Both boilers running at 90% rating. Both have no feedwater regulators. Without touching the feed valves, is it possible to get the both boiler's water level at 2-gages?

a. yes

b. no

71. If so, how? (question #70)
a. lower turbine speed
b. adjust firing rates
c. decrease vacuum on condenser
d. cannot be done. Answer is 'no' to question #70

72 You blow the column and glass and the column valve is worn and won't close all the way.

a. take boiler off the line

b. run boiler until convenience allows for repair

c. run boiler, but keep checking gage cocks

d. not so important if drain leaks

73. You blowdown the boiler and low water alarm activates, but you see 2-gages of water in glass (fire still on).

a. shut down right away

b. blow column and glass and try gage cocks

c. lower and raise fires to lower and raise water level to free the float in low water alarm. This is why the flame did not extinguish via LWCO.

74. A double acting nonreturn valve.

a. closes on tube or boiler failure

b. closes if main steam line breaks

c. can be manually opened and closed

d. all of the above

75. If boiler blowoff pipe breaks, which sequence is best?

a. kill fire only

b. kill fire. Open makeup water to receiver, then open bypass on feed water regulator

c. kill fire. Open bypass on feed water regulator then introduce makeup water to receiver

d. kill fire. Open makeup water to deareator

76. Staybolts in locomotive and vertical boilers support.

a. outside wrapper sheet

b. crown sheet

c. inside fire sheet

d. all of the above

77. Inside fusible plug.

a. 99% pure tin and must be removed from fire box

b. 99% pure tin and is replaced from internal steam and water section

c. 98% pure tin and replaced as mentioned in 'a' and 'b'

d. 98% pure babbitt and removed from inside fire box

78. Outside fusible plug.

a. 99% pure tin and must be removed from fire box

b. 99% pure tin and replaced from internal steam and water section

c. 98% pure tin and replaced as mentioned in 'a' and 'b'

d. 98% pure babbitt and removed from inside fire box

79. Three boilers on line with nonreturn valves and one feed pump. #2 boiler blows a tube.

a. open bypass on feed line wide to #2 and extinguish fires on #2

b. extinguish fires on all boilers

c. extinguish fires on #2, throttle bypass on feed regulator

d. perform 'c' but close bypass if #1 and #3 are losing water

80. To control superheat temperature, adjust

a. steam pressure and steam flow

b. gas baffle positioning and recirculation

c. adjust attemperator

d. all of the above

81. Superheated steam temperature can be controlled by

a. attemperation

b. movable gas baffles

c. steam pressure

d. all of the above

82. Pure natural gas is colorless odorless.

a. true

b. false

83. *To* boil-out a new boiler, a good formula is

a. 2 pounds of trisodium phosphate per 1,000 pounds of water

b. 2 pounds of caustic soda per 1,000 pounds of water

c. 2 pounds of soda ash per 1,000 pounds of water

d. mix all of the above

84. A main 'downcomer' tube is for positive convection flow to mud drum in water tube boilers.

a. true

b. false

85. Forced circulation in large watertube boilers is controlled by

a. steam pressure in drum

b. pumps

c. convection forces

d. density of water

86. Galvanic corrosion is caused by
a. dissimilar metal-to-metal contact
b. excessive feedwater chemicals
c. acid feedwater
d. alkaline feedwater

87. How would you stop galvanic corrosion inside a boiler steam drum?
a. lower chemical concentration
b. install zinc plates
c. decrease chemical concentration
d. adjust water PH

88. An electric resistance boiler has
a. low water cutout
b. gage glass
c. safety valve
d. set of electrodes
e. all of the above except 'a'

89. A high temperature hot water boiler (HTHW) does not require
a. gauge glass
b. water column
c. try cocks
d. all of the above

90. A HTHW boiler can explode if it starts steaming because relief valves cannot handle steam volume.
a. true
b. false

91. The water and steam drum on a supercritical once-through boiler is used for
a. drying steam continuously
b. starting up and steam separation
c. a main steam line
d. 'a' and 'c'

92. Circulation is accomplished in once-through boilers by
a. natural convection
b. steam flow
c pumps
d: 'a' and 'b'

93. Impurities in feedwater in a once-through boiler will
a. pass out with the steam
b. settle to mud drum
c. solidify in steam and water drum
d. 'b' and 'c'

94. What is a supercharged boiler?
a. boiler with two superheaters
b. boiler with a high furnace pressure
c. boiler burning volatile fuels in suspension
d. none of the above. No such boiler exists

95. If the flame trips out, what could you do to prevent thermal shock to a high capacity boiler?
a. close stop valve
b. close feed valve
c. stop forced or induced draft fans
d. all of the above

96. Some economizers have blowdown valves.
a. true
b. false

97. Modern high pressure boilers can tolerate high impurity concentrations in feedwater.
a. true
b. false

***98.* Temperature thermocouples found on or in a steam and water drum are most important to watch when**
a. starting boiler
b. shutting down boiler
c. normal running
d. 'a' and 'b'

99. Fuel oil pressure to a mechanical atomizing burner.
a. 25 p.s.i.
b. 45 p.s.i. .
c. 50 to 80 p.s.i.
d. 100 to 180 p.s.i.

100. Most efficient method to control the volume of air in a constant-speed draft fan.
a. controlling fan inlet
b. controlling fan outlet
c. controlling fan speed
d. recirculating air flow

101. A 100 horsepower boiler will generate ______lbs. of steam per hour from and at 212° F.
a. 3,450
b. 4,750
c. 100
d. 500

102. A flue gas sample should be taken not more than thirty feet from the opening of the stack outlet.
a. true
b. false

103. Pressure drop through the superheater has increased within the last 6 months. A new turbine with higher capacity is being installed. What must be done.
a. clean the boiler tubes
b. install more tubes
c. increase steam pressure
d. all of the above

104. Superheater temperature has declined.
a. raise fires to full capacity
b. burners running high CO readings
c. scale or soot on boiler tubes
d. all of the above

105. Electrochemical corrosion in a boiler is caused by
a. grease or oils
b. acid
c. 0_2
d. all of the above

106. Gagging a safety valve for a hydrostatic test could damage the valve so removing the valve and blanking the flange is preferred.
a. true
b. false

107. The pressure in the steam drum is higher than the superheater outlet.
a. true
b. false

108. Boiler and turbine running at 80% rated load. Boiler is in perfect tune with the turbine generator in relation to steam flow loads. All of a sudden the nonreturn valve on the boiler starts pounding.

a. boiler blew a tube
b. air in turbine hydraulics
c. flame failure
d. superheater safety valve opened

109. Starting a boiler with no superheater drains you should.

a. start as any conventional boiler
b. fire up quickly to create air and vapor flow through tubes
c. flood superheater with water
d. redirect gas passage

110. You're having a terrible time holding the water level in a boiler. It keeps dropping slowly out of sight and there is no low water cut-out, but just as the water went out of sight in the glass you noticed the steambound pump has just caught its prime. How could you check the boiler right now as to decide if to shut down the boilers fires or not too? (water column and glass are properly installed to code)

a. open then close drain to the gage glass
b. raise the fires, add feedwater to raise the water in gage glass
c. kill fire and restart to surge water level
d. close main stop valve partly to raise water in glass

111. A boiler with a surface blow operating continuously will not have to be blown down, only if surface blow used intermittently.

a. true
b. false

112. A flared tube is stronger than a beaded tube.

a. true
b. false

113. It is perfectly normal to see an HRT boiler's through-stays slightly bowed, as long as the through-stays nuts and washers are tight.

a. true
b. false

114. The reciprocating steam pump runs fine at 60% boiler rating, but when process steam demand on the boilers increases to 110% boiler capacity the pump always gets steam bound, but clears its self automatically when steam demand drops. This always seems to happen on your shift and is quite embarrassing. What is the problem to correct?

a. applying too much steam to closed heater
b. receiver water temperature
c. opening process steam valve to line capacity
d. pump packing is worn. Tighten packing gland nuts

115. Most undesirable effect when superheater safety valve lifts.

a. prime water into the superheater
b. valve may simmer upon reclosing
c. superheater will overheat
d. prime mover will slow down

116. Chemicals used to soften boiler feedwater (select 2 answers)

a. phosphate
b. sulfite
c. soda ash
d. lime

117. Four identical boilers on line. #1 is coal fired, #2 light oil, #3 heavy oil, #4 is gas. The prime mover is a 23,000 KW condensing turbine with its speed rapidly decelerating. Is it necessary to have all 4-boiler's steam pressure gages operating to determine the firing rates to compensate the turbine speed?

a. yes
b. no

118. Before firing heavy fuel oil you must circulate the oil to

a. strain the oil in filter screens
b. heat the oil
c. build up oil pressure
d. clean the burners of old oil

119. H_20 is entering the fuel oil to the burners and causing sputtering and flame-outs.

a. oil heater trap will not discharge
b. steam pressure to burner low
c. broken steam line in oil tank
d. leaking coil or tube in heater

120. Surface blow is properly located

a. in the steam and water drum
b. at steam drum water line
c. at mud drum water line
d. 6 inches from bottom of steam and water drum

121. Steam atomized burner is depositing oil on water wall tubes at 90% of boiler rating. It is best to cure this by

a. reducing primary air
b. reducing secondary air
c. installing a narrow pattern burner tip
d. reducing firing rate

122. Which burner is the most costly to operate?

a. rotary cup
b. steam atomized
c. air atomized
d. mechanical

123. Moderately hard water pertaining to scale forming impurities should read on a total hardness test

a. 300
b. 14
c. less than 6
d. 0-14

124. Draft pressure in a furnace combustion chamber for a balanced draft condition should read

a. atmospheric
b. +2 inches of water
c. -2 inches of water

d. -2 inches of mercury

125. Before performing safety valve capacity test.

a. lift valve handle to make sure it will lift
b. water level not more than 2 gages
c. install accurate pressure gage along with existing pressure gage
d. all of the above
e. 'b' and 'c' only

126. As steam load increases the radiant superheat temperature will rise.

a. true
b. false

127. Superheater drains are closed (placing boiler on line at 300 p.s.i.)

a. 150 p.s.i.
b. 225 p.s.i.
c. 300 p.s.i.

128. A pyrometer is used to measure

a. speed of turbine vibration cycles
b. temperature
c. alignment of coupling faces
d. steam flow
e. 'a' and turbine critical point resonance.

129. What would not cause high stack temperature?

a. high CO_2
b. broken gas baffle
c. excessive operation of soot blowers
d. overfiring burners

130. The draft gauge over the fires should read

a. negative pressure
b. positive pressure
c. no measurement taken here

131. To clean induction motor of grease use _____ with fuses removed and electric panel tagged and locked.

a. electrical cleaning solvent only
b. demineralized water
c. dielectric oil or grease
d. low pressure steam

132. To determine moisture content of steam, measure with a throttling calorimeter.

a. true
b. false

133. When purchasing a new boiler, which is least important to consider?

a. steam temperature and pressure
b. AC or DC to be generated by installed turbine
c. steam flow demand
d. fuel available

134. Turbulence is not desirable in the combustion process.

a. true
b. false

135. If you purchased a boiler that can generate 1,OOO°F superheated steam, could it efficiently drive a turbine / condenser package designed for 600°F steam?

a. yes
b. no

136 Select a reason for your answer to above question.

a. 600°F is more effective than 1,000°F
b. 1,OOO°F will damage turbine and overload condenser
c. boiler would be inefficient at turbine rating
d. 'b' and 'c'

137. Evaporating 34.5 pounds of water at 212°F in one hour under atmospheric pressure is equivalent to

a. 10,000 lbs./hr.
b. one boiler horsepower
c. 33,000 BTU's
d. 970 BTU's per second

138. To compensate a 12% increase in load on a oil fired boiler.

a. raise fuel pressure
b. change burner tip pattern
c. install a large forced draft fan and stack
d. 'b' and 'c'

139. You installed a new boiler five years ago running on gas fuel. Although fuel characteristics hasn't changed since you have just contacted the local utility and load has not changed in the past five years, you notice the flue gas CO_2 reading dropping steadily in the past years' time on a monthly basis test. What is wrong?

a. boiler water phosphate concentration rising
b. air leaking into breaching
c. gas pressure too high
d. induced draft fan decelerating or worn

140. When taking a boiler down for inspection, when should you completely drain the water?

a. boiler cold
b. boiler warm
c. boiler still steaming with stop valves open
d. boiler still steaming with stop valves closed

141. Reason for above answer.

a. impurities will solidify and fall to mud drum
b. so impurities will not precipitate and bake on metal
c. so steam pressure will force scale out of boiler
d. none of the above. Answer to question #140 is 'd'

142. Continuous blowdown line is normally used to take boiler water samples on the small firetube 150-horsepower boiler running at full capacity. Steam pressure is 70 p.s.i., but you walked onto your shift and were told the sample line is out of service for repair. Where will you take the sample from?

a. boiler blowdown tank
b. try-cock

c. feedwater or chemical feed injection line
d. steam drum vent

143. A watertube fails it will_____and when a firetube fails it will _____.
a. collapse
b. expand
c. compress
d. bend

144. The metal area of a head between firetubes is called a
a. ligament
b. tube sheet
c. stay
d. fill

145. Airheater recirculation dampers are put into operation.
a. at low loads
b. at high loads
c. continuously
d. when soot-blowing tubes

146. Reason for above answer.
a. to cool elements
b. to prevent corrosion
c. to create impingement effect
d. to gain a pressure drop

147. The pipe connection to the water column from the boiler should be angled so water will tend to flow into the boiler, but not sloped so water will tend to flow to the column.
a. true
b. false

148. Spalling of a furnace brickwork is caused by thermal shock.
a. true
b. false

149. A boiler gas pass baffle built as a one piece unit.
a. mortar and firebrick
b. laminated strips
c. sectional
d. Monolithic

150. Chemical analysis report on available water supply specifies 'temporary hardness' for the new power plant location site. You have selected to use 'D-type' water tube boilers for this plant. What feedwater treatment will be necessary?
a. Phosphate and sulfite
b. large capacity softeners
c. demineralizer
d. all are required

151. Match the appropriate answers in column I to column II. Example: 6 matches 'b'.

COLUMN - I

a. convection superheater =
b. forced draft = 6.
c. back pressure valve =
d. closed heater =
e. breaching air leaks =
f. radiant superheater =
g. horsepower =
h. open heater =
i. efficiency =
j. non-return valve =

COLUMN - II

1. check valve
2. output + input
3. primary superheater
4. secondary superheater
5. low CO_2 readings
6. primary air
7. 34.5 lbs. / hr
8. steam condensing
9. shell & tube heat exchanger
10. exhaust line protection

Chapter 12

Compressors Unlimited

The following are basic questions on gas compressors which are found in all power plants. Compressed air is used for operating automatic controls. To avoid major trouble always insure the air is free from moisture and lubricating oil or it will clog the tiny orifaces in the controls. When a device is acting erratic chances are air quality problems are developing. Blow down the air receiver and aftercooler daily. Keep compressor oil at recommended level. Never overfill crankcase. Make sure aftercooler is working properly and condensate trap is functioning on your shift.

1. A unloader device will most likely be found on a
a. reciprocating compressor
b. refrigeration absorption unit
c. centrifugal compressor
d. turbine driven rotary oil pump

2. Intercooler on a compressor
a. removes moisture or dew droplets only
b. condenses steam between stages in the air and removes condensate
c. cools crankcase oil
d. cools air so it can't back up into first stage

3. Aftercooler on a compressor.
a. removes moisture or dew droplets only
b. condenses steam between stages in the air and removes condensate
c . cools crankcase oil
d. cools air so it can't back-up into first stage

4. A compressed air receiver has a
a. relief valve
b. safety valve
c. safety relief valve
d. any of the above

5. On a compressed air intercooler or final discharge line from compressor you will find a
a. relief valve

b. safety valve
c. safety relief valve
d. any of the above

6. Compressor capacities are rated in
a. cubic feet per minute
b. pounds per square inch
c. feet per minute

7. A multistage intercooled and aftercooled compressor will
a. increase volumetric efficiency
b. decrease power consumption
c. reduces air temperature to its corresponding pressure

d. all of the above.
If air was backing out of intake intermittently on a reciprocating compressor.
a. piston rings worn
b. inlet valves not seating
c. bent connecting rod
d. all of the above

9. Oil in compressed air forms from (reciprocating)
a. oil vapor near the surrounding air intake
b. worn piston rings
c. overfilling crankcase
d. all of the above

10. For extended high volume CFM capacities, use a_____compressor.
a. reciprocating
b. centrifugal
c. rotary vane or axial
d. 'b' or 'c'
e. all of the above

11. The diffuser in a centrifugal compressor is to
a. accelerate air flow
b. absorb heat
c. convert air velocity to pressure
d. diffuse air to reduce noise

12. The most efficient compressor.
a. reciprocating
b. centrifugal
c. rotary screw
d. rotary vane
e. axial

13. For oil free air, use a centrifugal compressor.
a. true
b. false

14. Some rotary screw compressors require oil in the air for lubrication.
a. true
b. false

15. The rotary vane compressor can be started 'unloaded'.
a. true
b. false

16. Rotary vanes must be lubricated by
a. air
b. water
c. oil

17. Rotary vanes contact cylinder surface on start up.
a. yes
b. no

18. Axial compressor.
a. horizontal opposed pistons in one cylinder
b. utilizing turbine type blades on rotor and casing
c. multistage positive displacement compressor
d. all of the above

19. Vanes contact cylinder surface in proportion to
a. temperature
b. centrifugal force
c. oil pressure
d. air flow volume through machine

20. Unloading a compressor is accomplished by
a. intercooled recirculation
b. opening intake valves
c. intake air throttling
d. all of the above

21. You will find a _____on the discharge line on a rotary vane compressor.
a. oil injection system
b. humidifier
c. oil separator
d. 'a' and 'c'

22. A safety or relief valve must be installed on the discharge line between the compressor and the first stop valve.
a. true
b. false

23. If oil with a low flash point is used in a reciprocating compressor and intercooler or aftercooler fails, a fire could result depending on heat of compression temperature.
a. true
b. false

24. The receiver stores air so compressors will not run continuously, and demand can be satisfied without an immediate pressure drop throughout the system.
a. true
b. false

25. Centrifugal and axial compressors are dynamic class compressors.
a. true
b. false

26. A dynamic compressor imparts energy to a gas by the use of blades building up velocity and pressure.
a. true
b. false

27. To remove oil from the air from a centrifugal compressor install a_____on discharge line.
a. oil separator
b. trap
c. oil filter
d. no oil in air from centrifugal

28. A pressure control switch takes the place of a
a. relief valve
b. unloader
c. check valve
d. none of the above

29. The unloader
a. prevents overloading the driver
b. permits compressor to start at low speed
c. relieves the cylinders of compression
d. 'c' and 'a'

30. Relief valve can be located on
a. receiver
b. discharge line

c. between stages
d. inter or aftercoolers
e. all of the above

31. Air compressor that works on both sides of a piston.
a. single acting
b. double acting
c. 2 stage
d. 'a' and 'c'

32. A compressor can be driven by
a. electric motors
b. steam turbines
c. diesel engines
d. gasoline engines
e. all of the above

33. The air compressors available are
a. reciprocating
b. rotary vane
c. axial
d. rotary screw
e. all of the above

34. Intercooler is used to
a. increase volumetric efficiency
b. cool oil
c. increase compression ratio
d. remove oil and 'b'

35. Finger valves or reed valves are strips of metal covering air inlet and outlet ports and are fully automatic in operation.
a. true
b. false

36. Compressor with finger valves.
a. reciprocating
b. centrifugal
c. rotary
d. axial

37. If oil level is too high in a reciprocating compressor
a. oil could damage equipment downstream
b. piston rings will wear out cylinder walls
c. receiver capacity could be reduced
d. 'a' and 'c'

38. Relief valve started cycling on a reciprocating compressor's receiver.
a. piston rings broken or worn
b. coolant to aftercooler failed
c. pressure control switch malfunction
d. cylinder head gasket blown

39. Centrifugal compressors need no intercoolers.
a. true
b. false

40. Aftercooler's main function.
a. condense non-condensable gasses
b. remove oil
c. condense steam from air
d. all of the above

41. To remove oil in large quantities from air, install a
a. intercooler
b. aftercooler
c. separator that changes air direction
d. any of the above

Chapter 13

Steam Engines Unlimited

1. Steam engine and steam reciprocating pump in operation. Oil entering the boiler.
a. cylinder oil not properly metered
b. grease extractors failed or clogged up
c. cylinder oil reservoir filled with light engine oil
d. all of the above

2. A steam engine valve has 'cutoff'
a. for expansion, efficiency, governing speed of engine, piston cushioning
b. so pressure will not rise too high in cylinder
c. so condenser will not be overloaded with exhaust steam
d. all of the above

3. Superheated steam can be used in steam engines.
a. true
b. false

4. What is the limiting factor on using superheated steam in steam engines?
a. no limiting factor. Superheat can be used at any temperature
b. superheated steam has too much power-bending piston rods
c. cylinder oil will break down
d. superheated steam cannot be used. Piston will freeze in cylinder due to cylinder contraction

5. Oil used in steam the end of engines and pumps.
a. gear oil
b. mineral oil
c. cylinder oil
d. engine oil

6. The counterbore at the cylinder ends
a. prevent shoulders and resultant packing stuffing box leaks
b. to help insert new piston or piston rings or cups
c. to measure for a rebore of the cylinder
d. all of the above

7. Slide valve steam engines have 'lost motion'.
a. true
b. false
c. some do and some don't

8 . Valve is not used in steam engines or pumps.
a. B-type valve
b. D-type valve
c. poppet valve
d. Corliss valve
e. butterfly valve

9. The low pressure cylinder in a compound engine is
a. smaller
b. larger
c. same size as high pressure cylinder

10. Tram a steam engine to find
a. half stroke of piston

b. dead centers
c. valve cutoff point
d. crankshaft starting angle

11. Admission of steam begins
a. at dead center
b. after dead center
c. before dead center
d. 10° of crankshaft angle

12. To reverse a engine's direction of rotation (engine running over)
a. place eccentric 90° plus lap and lead behind the crank
b. place eccentric 90° plus lap and lead ahead of the crank
c. tram engine then perform 'a'

13. A uniflo engine has
a. no exhaust valves
b. has 2 exhaust valves
c. has no admission valves
d. uniform steam cylinder temperature and pressure

14. Method to reduce condensation in engines.
a. install a condenser
b. steam jacketing
c. superheating steam
d. 'b' and 'c'

15. Valve not compatible with superheated steam
a. slide valve
b. poppet valve
c. Corliss valve

16. Slide valve engine with too little lost motion
a. piston will use too much steam
b. piston will short stroke
c. piston will strike cylinder heads
d. no lost motion in steam engine

17. Steam engines have a overspeed trip that senses Rpm's.
a. true
b. false

18. Condensation and re-evaporation in steam engines
a. beneficial to efficiency
b. undesirable, but not much can be done about it
c. re-evaporation does not exist in steam engines

19. Steam jacketing refers to
a. steam blanketing of cylinder exterior
b. quick opening and closing of throttle
c. feedwater heaters
d. all of the above

20. Can a counterflow slide valve engine have variable cutoff?
a. yes
b. no

21. A hydrostatic lubricator works by
a. steam pressure
b. displacement of oil by condensate
c. gravity
d. all of the above

22. A force-feed lubricator functions on the principles of
a. pistons or rams
b. hydraulic force
c. mechanical energy from engine piston rod or crosshead
d. all of the above

23. Can exhaust steam from engines be used in open heaters?
a. yes - no precautions are necessary
b. no
c. some engines yes - some no
d. yes - if oil separator is installed on exhaust steam line

24. Cross-compound engine. What is placed between the high and low pressure cylinders?
a. back pressure valve
b. receiver
c. relief valve
d. check valve

25. The eccentric on a engine operates
a. governor
b. valve gear
c. tachometer
d. all of the oil systems

26. The direction of rotation of flywheel on a engine determines if it is running over or under?
a. true
b. false

27. The angle of advance on a steam engine refers to
a. governor speed selection
b. direction of flywheel rotation
c. valve setting
d. piston stroke length

28. What is the angle of advance?
a. 90° plus 10° lead
b 90° **plus 10°** lap
c: 90° plus 10° lap plus 10° lead
d. There is no angle of advance on steam engines

29. The flywheels purpose is to
a. absorb and release energy
b. soften harsh engine forces
c. bring pistons past dead centers
d. all of the above

30. Dashpots on steam engines are used to
a. control piston speed
b. feed oil to cylinders
c. dampen the valve gear sensitivity
d. all of the above

31. A steam engine with a throttling governor will not require
a. lubrication
b. belts
c. variable valve cutoff
d. overspeed protection

32. Large electrical cables are always stranded because they are
a. lighter
b. stronger
c. flexible
d. low in electrical resistance

Chapter 14

Pumps Unlimited

1. Absolute pressure
a. 30" of water
b. atmospheric pressure plus gauge pressure
c. 29.9" mercury
d. **'a'** plus 'c'

2. After packing the cylinders of a reciprocating pump, you start it and suddenly it stops after five minutes. What is wrong?
a. exhaust valve closed
b. high water in boiler. No demand on pump
c. packing rings too tight binding piston in cylinder
d. oil in exhaust steam

3. Compression or cushion valves on steam reciprocating pumps
a. increase capacity of pump
b. increase speed
c. lower speed
d. prevent pistons from striking cylinder heads

4. Reciprocating steam pumps for boiler feed use, have
a. governors
b. high speed trip
c. low speed alarm
d. all of the above
e. none of the above

5. Power pump
a. large single piston
b. turbine driven centrifugal pump
c. multiple pistons or plungers on single crank shaft
d. multiple crankshafts, each driving a centrifugal pump

6. Plunger pump features
a. packing installed on piston
b. packing installed on plunger
c. packing installed in stuffing boxes
d. none of the above

7. Rotary pump utilizes
a. pistons
b. screws
c. impellers
d. gears

8. Rotary screw pump

a. screw rotating in housing or with another screw
b. screw plunger thru packing
c. reciprocating screw with pistons attached
d. any of the above

9. Which cylinder on duplex pump is the longest in length?
a. steam end
b. water end
c. both the same

10. The smallest pipeline on a pump is
a. suction line
b. discharge line
c. both the same

11. Running a boiler with a centrifugal feed pump. Boiler at 100% capacity, pump running at maximum, receiver has 21" of water in glass. Pump keeps getting air bound.
a. water too hot
b. water too cold
c. vortex in receiver
d. if seals and joints don't leak this can't happen

12. What lubricates piston rod on 'water end' of a duplex pump?
a. steam
b. water
c. oil
d. packing

13. All steam driven reciprocating pumps develop compression in the 'steam end'.
a. true
b. false

14. A duplex pump with no lost motion would
a. stop running
b. never pause at end of stroke
c. overspeed the pump and pistons will strike cylinder heads
d. this pump has no valve adjustments

***15.* Steam reciprocating pump for boiler feed. Which piston is the largest?**
a. water
b. steam
c. both the same size

16. Lost motion in steam reciprocating pump valves.
a. pump can't overspeed
b. pistons can't bump cylinder heads
c. allow pistons to come to rest
d. steam pumps of this kind have no lost motion

17. Could you say a duplex steam reciprocating pump is basically a counterflow type engine?
a. yes
b. no

18. A steam reciprocating pump starts racing and bumping heads.
a. steam bound
b. air bound
c. compression valves closed
d. drip valves left open

19. In a steam reciprocating slide valve pump. Slide valve covers the steam port and piston is covering the exhaust port. What condition is this event called?
a. dead center
b. admission
c. exhaust
d. compression

20. To prime a duplex steam pump
a. fill discharge pipe with water
b. fill discharge chamber with water
c. fill suction line and suction chamber or cylinder with water
d. duplex pumps need no priming because they are positive displacement type pumps

21. Centrifugal pumps are positive displacement pumps.
a. true
b. false

22. Too much lost motion in valve gear on duplex pump.
a. piston short-strokes
b. piston strikes head
c. none of the above

23. Too little lost motion in valve gear on duplex pump.
a. piston short strokes
b. piston strikes head
c. none of the above

24. Duplex piston pump is a positive displacement pump.
a. true
b. false

25. Rotary pump is positive displacement.
a. true
b. false

26. Positive displacement pumps never get air bound.
a. true
b. false

27. Centrifugal pump that spins liquid in many individual volutes within one impeller rotation.
a. regenerative pump
b. turbine pump
c. 2-stage centrifugal
d. 'a' and 'b'

28. A multiple stage centrifugal pump must have
a. Two main suction line connections
b. Two main discharge line connections
c. Two or more impellers
d. all of the above

29. A reciprocating pump with no piston.
a. rotary
b. screw
c. diaphragm
d. jet

30. Chemical pumps used to treat boiler water are called
a. metering
b. proportioning
c. adjustable constant feed
d. any of the above

31. To control discharge pressure on a motor driven centrifugal pump
a. close down on discharge or recirculation valves
b. regulate speed by lowering Hertz
c. close down on suction valve
d. all of the above

32. Turbine driven feed pump pressure is erratic. Check the
a. feedwater regulator to boiler
b. governor sensitivity control
c. cooling water to bearings
d. governor oil level
e. all of the above

33. When shutting down a pump that may experience freezing temperatures
a. keep pump full of water
b. drain water completely
c. keep pump warm with electrical heat tape
d. 'b' or 'c'
e. all of the above

34. The relief valve on the discharge line from a centrifugal pump is to prevent heat build-up, and maintain flow within the pump than to solely protect pump from over pressurizing the casing.
a. true
b. false

35. Classes of centrifugal pumps.
a. axial-flow
b. horizontal or vertical
c. mixed-flow
d. radial or centrifugal
e. double or single suction
f. single or multistage
g. all of the above

36. NPSH
a new pump start hot
b: net positive suction heat
c. net positive suction head
d. never pump solids hard
e. National Pump Society of Hydraulics

37. Multistage centrifugal pumps actually have a heating effect on the water, and can be desirable for boiler feed service, within limits.
a. true
b. false

38. To increase the output of a centrifugal pump.
a. speed up rotation
b. increase suction pipe area
c. install a recirculation line
d. all of the above

39. The highest level a pump could lift under an ideal vacuum.
a. 10.5 ft.
b. 25 ft.
c. 33.9 ft.
d. 48.3 ft.

40. The air chamber on the discharge line of a boiler feed pump is to collect water, but restrict the flow to smooth pulsation's created by piston irregularities in stroke speeds.
a. true
b. false

41. Volute chamber or diffuser ring in a centrifugal pump acts to
a. prevent shaft wear
b. change velocity to pressure
c. prevent stage slips (leakage)
d. cushion the flow

42. Eight-stage centrifugal pump will have a_____which a small two-stage pump may not have.
a. impeller shrunk on shaft
b. thrust bearing
c. volute chamber
d. governor

43. Can be used to feed a boiler (select four answers).
a. return trap

b. city water line
c. separating trap
d. ejector
e. injector
f. Inspirator

Chapter 15

Turbines Unlimited

1. Steam turbine.
a. utilizing expanding steam in diaphragms to produce back pressure on blades
b. utilizing kinetic energy in expanding steam to produce mechanical energy
c. utilizing impulse effects in diaphragm

2. Reaction turbine.
a. steam striking blades on angle
b. steam reversing direction
c. steam striking blades at zero angle
d. reacts to steam pressure and velocity in diaphragm

3. Impulse turbine.
a. steam strikes blades on angle
b. steam reversing direction
c. steam strikes blades at zero angle
d. reacts to steam pressure and velocity in diaphragm

4. Gas turbine.
a. burns natural gas for fuel
b. uses steam for fuel
c. uses any high grade volatile fuel in fluid form
d. air or steam for fuel

5. Noncondensing turbine.
a. back pressure turbine
b. condensation takes place not in turbine, but in exhaust line
c. when condenser is shut down
d. is a compressed air turbine

6. Condensing turbine.
a. back pressure turbine
b. condensation in exhaust system
c. condensation begins in last stage
d. any turbine with condenser in low pressure exhaust line

7. Back pressure turbine.
a. condensing turbine
b. noncondensing turbine
c. low vacuum on exhaust line
d. opposing pressure on governor throttle valve

8. Opposed flow turbine.
a. steam flowing in one direction
b. steam flowing in two directions
c. steam flows in three directions
d. steam flowing opposite direction of rotating blades

9. Extraction turbine.
a. steam taken from exhaust line to process work
b. steam looped from second stage back to first stage
c. steam taken from any stage
d. all of the above

10. Regenerative turbine.
a. complex multistage impulse turbine
b. simple single wheel reaction turbine
c. extraction reheat turbine
d. steam from second stage recirculated to first stage

11. Main bearings support
a. thrust pad on rotor
b. rotor
c. casing
d. governor

12. Labyrinth seals are located
a. on the rotor
b. in the casing as shaft exits
c. between the stages only
d. all of the above

13. Thrust bearing.
a. horizontally lines up turbine clearances
b. is not adjustable and must be replaced as clearance increases
c. produces thrust against main bearings for stability
d. has a pad and roller bearings to assist thrust

14. Sealing strips.
a. direct steam into blading only
b. protect bearing oil from entering the turbine
c. reduce steam leakage and short circuiting around blades
d. seals the casing halves

15. Diaphragm.
a. steel of alloy metals which blades are dove-tailed to
b. nozzle plate of each beginning stage
c. controls governor throttle valves
d. solid steel plate between stages

16. Admission valves are located in a nonextraction turbine
a. at every stage
b. insteam chest
c. in extraction section
d. in steam line

17. Sentinel relief valve
a. protects against high casing pressure
b. warns of high casing pressure
c. vacuum breaker for casing
d. is located on exhaust line to protect high back pressure

18. Overspeed trip is
a. electrically operated
b. manually operated
c. hydraulic oil operated
d. all of the above

19. Oil relay governor.
a. pressure operated from steam
b. works off of main bearing oil pump only
c. shaft driven oil pump and oil pressure operated
d. relays oil to bearings to control friction therefore speed

20. Centrifugal governor
a. operates with oil
b. operates in oil bath
c. does not need oil only spring and weights
d. none of the above

21. Drains are located
a. lowest points in casing at each stage
b. one drain on lower casing and one at each main bearing
c. steam chest, labyrinth seals, thrust bearing, carbon ring seals
d. 'a' and 'c' and thrust bearing

22. Blading shrouds located
a. in dovetail and wheel connection of blades
b. in casing
c. on blade ends
d. all of the above

23. Blading shrouds
a. strengthen blading and help guide steam and acts as a flywheel
b. hold blades to the wheel or disk
c. direct steam out of the blades
d. protect blades from wet steam conditions

24. Babbitt is made of
a. copper and lead
b. aluminum and tin
c. tin, antimony and copper
d. molybdenum oxide

25. Main Babbitt bearings when made are
a. forged
b. stamped
c. poured
d. laminated

26. The turbine cylinder.
a. casing including all inside area
b. area of blading and casing steam path
c. nozzle and diaphragm plate
d. steam chest

27. Blades when installed are
a. dovetailed and welded
b. dovetailed, pressed and pinned with a shim
c. dovetailed and strung with wire to rotor disk
d. dovetailed and peaned with a special hammer

28. Extraction valves are
a. slotted disks
b. gate valves single or double split wedge type valves
c. globe valves for control
d. check valves

29. All governor valves are
a. double seated
b. single seated
c. balanced
d. lubricated

30. Overspeeding causes damage from
a. vibration
b. oil pump capacity too small for this higher speed damages bearings
c. high steam pressure
d. centrifugal force

31. The initial stage.
a. steam chest
b. first stage
c. second stage
d. last stage

32. Moisture in steam causes
a. pitting of blades
b. vibration and friction, reducing efficiency
c. erosion of rotating, stationary blades, labyrinth seals and main bearings
d. all of the above

33. Hunting.
a. governor valve controlling speed
b. valve position searching for a constant
c. steam surging in steam chest
d. all of the above

34. High back pressure.
a. noncondensing turbine
b. condensing turbine
c. exhaust over 15 p.s.i.
d. exhaust below 15 p.s.i.

35. Low back pressure is
a. noncondensing turbine
b. condensing turbine
c. exhaust over 15 p.s.i.
d. exhaust p.s.i. below 15

36. Turbine wheels or disks are
a. shrunk on the rotor, forged or keyed to rotor
b. threaded to rotor
c. bolted to rotor
d. must be welded only

37. Cooling water to a turbine is used to cool
a. rotor
b. labyrinth seals
c. carbon rings
d. bearings via oil coolers

38. Carbon rings.
a. seal steam and air leakage
b. used to lubricate shaft
c. prevents oil entering turbine
d. located at each diaphragm

39. Auxiliary oil pumps are used to
a. energize governor controls and lubricate bearings
b. lubricate bearings
c. drain oil from casing
d. 'b' and 'c'

40. Turbine casing halves are
a. gasketed to stop steam leaks
b. sealed with compound and gaskets
c. metal-to-metal fit without a gasket, but may use liquid sealing compound to seal joint from leaks
d. dovetailed and pinned together

41. Pillow blocks.
a. main bearing and journal supports
b. thrust bearing material
c. the turbine foundation
d. prevents surging in oil case

42. Power is developed initially
a. in the rotor
b. in the diaphragm
c. in the rotating stationary elements
d. in the last stage in a condensing turbine due to large blades

43. What causes thrust on the thrust bearing of a condensing turbine?
a. centrifugal force
b. vacuum in the condenser
c. pressure differential across turbine
d. all of the above

44. What type turbine requires the largest thrust bearing?
a. impulse
b. reaction
c. makes no difference unless it's a superheated turbine
d. both require the same size

45. All turbines must have overspeed protection.
a. true
b. false
c. true - only if in electric generation
d. false - if turbine is running reducing gears

46. Turbines are better than steam engines.
a. oil free exhaust steam and less moving parts to wear
b. high speed and high power output
c. efficiency and smaller size
d. all of the above

47. Before starting a condensing turbine first
a. start condenser
b. roll the turbine with the turning gear
c. open admission valves to increase back pressure
d. all of the above

48. When shutting down turbine or engine
a. close steam admission valves quickly
b. close steam admission valves very slowly
c. open casing drips first
d. perform 'c' then 'a'

49. Turning gear on a turbine.
a. refers to a turbine driven gear reducing system
b. keeps oil pump at full capacity
c. prevents rotor warping
d. makes it easier to start turbine by overcoming inertia

50. Decreasing back pressure on a turbine or engine
a. decreases efficiency
b. increases efficiency
c. has no effect
d. could have damaging affects

51. A turbine condenser should have
a. vent
b. vacuum breaker
c. atmospheric relief valve
d. safety valve with blow back

52. A noncondensing turbine exhaust line you will find
a. safety valve
b. relief valve
c. check valve
d. nonreturn valve and safety valve

53. Starting a steam turbine too quickly will cause
a. cracks in metal
b. blade warping
c. rotor bowing
d. all of the above

54. Water jackets on a turbine are to
a. cool rotor
b. cool oil
c. keep seals tight
d. prevent overheating of casing

55. If a condensing turbine is losing power, but condenser is working fine. What is wrong?
a. superheat temperature falling
b. oil temperature too hot
c. extraction valves stuck closed
d. 'c' but extraction to closed heater is open

56. Condensing steam turbine is consuming more steam than normal.
a. labyrinth seals worn-out
b. superheat temperature falling
c. casing drains left open
d. all of the above

57. Can a centrifugal spring type flyball governor be used to generate alternating current electricity?
a. yes
b. no

58. The auxiliary oil pump is for
a. changing oil on turbine only
b. circulating oil to labyrinth seals
c. controlling only the governor when turbine is running with load
d. starting and stopping turbine

59. A turbine with one stage and multiple reversing blades is a

a. compounded multiple stage
b. opposed flow turbine
c. compounded turbine

60. A condensing turbine runs efficient in practice at
a. 29" of mercury
b. 27" of mercury
c. 35" of mercury
d. steam temperature twice the saturation temperature

61. Boiler superheater tube failure can be caused by.
a. broken gas baffle
b. broken separator
c. overfiring
d. all of the above

62. To increase turbine horsepower.
a. run oil temperature 20°F below normal then perform 'c'
b. partly close exhaust valve
c. put another boiler on the line
d. install a condenser, use superheated steam, raise steam pressure to turbine

63. Labyrinth seal.
a. female grooves on rotor - male intrusions within female grooves with no physical contact
b. female grooves on rotor - male intrusions within female grooves with metal-to-metal contact
c. split carbon rings retained by springs to maintain rotor contact

64. The last stage in a condensing turbine, blades are
a. larger
b. smaller
c. same size as first stage
d. larger in a impulse turbine

65. The needle valve adjustment on the oil relay governor
a. adjusts shaft speed
b. adjusts governor sensitivity
c. adjusts overspeed trip setting
d. sets speed droop

66. Turbine casing drains are piped directly to a blowoff tank.
How do you know when to close the drain valves?
a. see steam from blowdown tank vent
b. feel drains
c. listen to drains
d. all of the above

67. Is it possible for a turbine to rotate backwards when starting up?
a. yes
b. no

68. Speed droop on a turbine governor is used to
a. adjust shaft speed only
b. balance loadings
c. synchronize generators
d. only used on steam engines

69. Condenser loses vacuum.
a. turbine will slow down to 80% rated speed
b. bearing oil will overheat
c. atmospheric relief valve will open
d. turbine will stall

70. Air ejectors are always operating on a condenser due to
a. noncondesable gasses in boiler feedwater
b. condenser tube leaks
c. leaking labyrinth seal
d. any of the above

71. What is load?
a. weight of the prime mover
b. workload produced by prime mover = load
c. accumulation of steam in the steam chest plus 'a'
d. 'a' and 'b'

72. Starting steam turbine or engine with drains and drips piped directly into a condenser. When would you close the drips?
a. 20 seconds after opening throttle
b. when turbine is at full speed
c. when you see steam from drips
d. drips are hot and vacuum increases in condenser

73. High vacuum in a condenser.
a. too much cooling water
b. vacuum pumps not working
c. low water in hot well
d. is very desirable

74. Back pressure on a noncondensing turbine.
a. is designed to run this way
b. hazardous situation above 14.7 p.s.i. absolute pressure in exhaust line
c. condenser failure
d. too little load on turbine

75. A turbine in good mechanical order starts vibrating. Check for
a. superheated steam too hot
b. steam separator malfunction
c. back pressure too low
d. low water in boiler

76. A flywheel on engine or shrouding on turbine.
a. absorbs and releases energy due to inertia effects
b. holds the wheel or blades together
c. produces kinetic energy due to centrifugal force
d. 'b' is not correct
a. 'a' and 'b'

77. When starting a engine or turbine with casing drains open.
a. open throttle slowly and smoothly
b. crack open throttle, then open quickly, then back it off

c. crack open throttle, wait one minute, then open very slowly
d. makes no difference

78. 'Critical points' in a engine or turbine when starting up.
a. governor taking control
b. harmonic vibrations forming
c. opening the throttle on a idle engine or turbine too slowly
d. closing the casing drains

79. What do you do when prime mover comes to acceptable critical points when starting up?
a. speed up turbine-open throttle valve slowly
b. slow down turbine-closing throttle valve
c. let throttle stay in position and open drains
d. close drains and trip overspeed to shut down

80. The spring on a flyball governor tends to
a. open the throttle
b. close the throttle
c. neutralize steam forces on throttle and centrifugal force of flyballs
d. operate a tripping mechanism

81. The low pressure side of a steam turbine is
a. smaller
b. larger
c. same size as high pressure size

82. A turbine shut down and not rotating can be destroyed by
a. ironrust on blades
b. gravity
c. no oil on bearings
d. no damage will result

83. Exhaust steam is_____in a noncondensing turbine.
a. wet
b. dry
c. vapor
d. corrosive

84. The mineral that causes the most problems in turbines introduced by the feedwater in boiler.
a. calcium
b. iron
c. silica
d. none of the above can evaporate and condense in turbine

85. Turbine trips out.
a. overspeeding
b. oil pump failure
c. cooling water failure or condenser vacuum loss
d. any of the above could trip it

86. Turbine keeps tripping on overspeed. Cooling water, oil, condenser is okay.
a. initial steam pressure too high
b. faulty governor
c. overspeed trip out of adjustment
d. any of the above

87. To adjust back pressure on a condensing turbine.
a. regulate condenser water flow
b. adjust cooling tower fan loads
c. adjust turbine loading
d. all of the above

88. When shutting down a large multistage turbine.
a. trip overspeed by hand
b. reduce speed slowly
c. shut down condenser, then perform 'b'
d. 'c' then 'a'

89. If you heard pinging noises from a turbine and boilers were not priming. What would you do?
a. open casing drains
b. reduce load
c. increase load, perform 'a'
d. blow down boilers

90. If turbines are heat engines, why will it run on cold ambient compressed air?
a. turbines do not require heat, therefore it will run
b. the air contains heat in relation to its liquid state
c. turbines are not heat engines, only steam engines are
d. Air will not run the turbine
e. 'a' and 'c'

91. A steam turbine can be driven at lower than rated speeds and maintain design efficiencies.
a. true
b. false

92. What would you do if the boiler feed pump's steam turbine bearings started to make noise?
a. decrease speed
b. take pump out of service
c. increase oil pressure
d. increase the pump speed

93. Abnormally high oil pressure on bearings will cause vibration.
a. true
b. false

94. Air in the oil of a hydraulic governor will cause hunting.
a. true
b. false

95. Excessively tight carbon rings will cause
a. shaft scoring
b. vibration
c. carbon ring damage
d. all of the above

96. Steam and exhaust piping strains on turbine will throw shaft out of alignment.
a. true
b. false

97. Vibration can be caused by
a. coupling misalignment or high oil pressure to bearings
b. tight or contaminated carbon rings or wet steam from boilers
c. worn or broken buckets
d. shroud damage or boiler compound accumulation
e. all of the above

98. It is advisable to warm up the casing by cracking open the steam valve with rotor not in motion.

a. true
b. false

99. What is a rotor locating bearing?

a. thrust bearing
b. main bearing
c. a bushing or sleeve
d. all of the above

100. 'Jacking gear' is another term used for turning gear on a turbine.

a. true
b. false

101. Steam nozzle hand valves are to

a. be either wide open or closed in operation to prevent wear
b. carry full-rated load at reduced steam pressure
c. admit steam to nozzles
d. all of the above

102. Overspeed trip functions, but admission steam valves did not close.

a. bent valve stem
b. worn valve seat or boiler compound accumulation
c. bushings or stem bearing worn
d. improperly adjusted steam valve mechanism
e. all of the above

103. Can you run an electric AC generator turbine unit on manual speed control if the automatic controls have failed to maintain constant speed?

a. yes
b. no

104. Another name for sealing strips.

a. packing seals
b. glands
c. spill strips
d. all of the above

105. Pivot type thrust bearings and tilting pad journal bearings can tolerate some misalignment.

a. true
b. false

106. A turbine's AC generator can be cooled by

a. air
b. hydrogen
c. 0_2
d. H_20
e. 'a' or 'b' or 'd'

107. Thrust bearing damaged.

a. dirty oil or no oil
b. boiler priming
c. applying full load too quick
d. overloading
e. any of the above

108. Hydrogen gas is explosive with oxygen when ignited.

a. true
b. false, it burns very slow

109. Water intrusion damage in a extraction turbine other than boiler priming could come from

a. feed water heater
b. attemperators
c. condenser
d. any of the above

110. Hydrogen gas is used in AC generators because

a. reduces friction
b. cools windings
c. will not leak from generator
d. 'a' and 'b'
e. cannot be used in AC generators

111. By closing the extraction lines to feed water heaters, the turbine will develop more useable power, but heat rate cycle will drop.

a. true
b. false

112. The best place to measure vacuum on a condensing turbine.

a. condenser water level
b. turbine exhaust hood
c. air ejector inlet

113. Wet steam to a turbine will

a. erode elements
b. cause vibration
c. increase steam consumption
d. all of the above

114. How often would you internally inspect your turbine?

a. once a month
b. once a year
c. once every three years
d. once every five years

115. To avoid distortion and rubbing in a turbine on startup, thermocouples should be placed on the

a. rotor
b. casing
c. diaphragm
d. all of the above

116. A labyrinth seal do not need to have male and female groves, but may have only male extensions extending to and surrounding the rotor shaft.

a. true
b. false

117. Would you run your turbine if sealing steam to the seals failed letting air into the condenser, but vacuum was holding within acceptable limits?

a. yes
b. no

118. To find the answer to the above question #117. what will cold air do to a hot rotor?

a. cause thermal shock
b. cracks will form in rotor
c. all of the above

119. Turbines utilize a shaft governor.

a. true
b. false, only steam engines do.

120. Turbines can utilize a throttling governor.

a. true
b. false, only steam engines do

121. Condensing turbine's atmospheric relief valve fails to open. This will

a. destroy condenser
b. burst turbine exhaust hood
c. damage turbine
d. all of the above
e. none of the above

122. High water in deareator. Turbine exhaust line is tied to the Da. Traps are handling the overflow on exhaust line and the turbine is at full load. Suddenly the turbine explodes one hour later with the same condition of high water in DA and steam traps discharging. In fact, the make-up to DA is off and condensate return pump is bypassed to the sewer. What caused this?

a. turbine extraction valve opened
b. turbine load reduced
c. turbine load increased

d. turbine cannot be damaged

123. A bleeder turbine.
a. opposed flow
b. non-condensing
c. extraction
d. induction

124. One-500 kw AC generator on the same bus as a 1,200kw AC generator. To properly balance both turbines to their maximum rated loads
a. adjust governor speed droop
b. overspeed 500kw turbine 10%
c. increase voltage on 500kw generator 10%
d. reduce 1,200kw generator to 500kw

125. Another name for a velocity stage is a 'Rateau'.
a. true
b. false

126. Two steam turbines located on a single rotor is not likely a
a. cross-compound
b. tandem compound
c. compounded induction
d. tandem cross-compound

127. Turbines are used today because
a. of very high starting torque
b. no reducing gears necessary
c. can be run at low speed
d. none of the above

128. A typical hydraulic turbine governor utilizes
a. springs
b. fluid
c. pneumatic forces
d. all of the above

129. A centrifugal governor utilizes
a. springs
b. oil
c. weights balanced by oil p.s.i.
d. pneumatic forces

130. It is possible to use a flyball governor on mechanical drive turbines.
a. true
b. false

131. Condensing turbines run efficiently at___inches of mercury.
a. 14.7
b. 18.7
c. 25
d. 29.2

132. A twenty-one stage impulse turbine's nonrotating elements will
a. expand the steam
b. compress the steam
c. redirect the steam
d. prevent short circuits

133. A turbine with nozzle hand valves open will
a. increase power output
b. increase initial steam pressure
c. increase steam chest pressure
d. overspeed the turbine
e. all of the above

134. Before starting a new power plant with a turbine you must_______to protect this turbine from serious damage.
a. open casing drains and lubricate governor linkage
b. air or steam-blow main steam and extraction steam lines
c. boil-out the boiler with superheater drains open
d. roll turbine with steam when performing 'c'

135. Match the appropriate questions in column I to answers in column II. Example: a = 4.

COLUMN - I

a. carbon ring = 4
b. main bearing =
c. diaphragm =
d. thrust bearing =
e. vibration =
f. rotor location =
g. speed droop =
h. critical point =
i. tachometer =
j. reversing element =

COLUMN - II

1. vertical, horizontal rotor stability
2. axial alignment
3. beginning of a stage
4. shaft seal
5. harmonic vibration
6. governor
7. rotor imbalance
8. thrust bearing
9. stationary blade
10. revolutions per minute

136. A turbine's steam condenser will
a. produce vacuum
b. condense steam
c. deaerate
d. all of the above
e. 'a' and 'b'

Chapter 16

Physics

1. Zero degrees centigrade.
a. freezing point of water
b. boiling point of water
c. there is no zero on this scale

2. Water will boil at a lower temperature
a. under pressure
b. under vacuum
c. with 100 p.s.i. of nitrogen acting on surface
d. in low elevation deserts

3. Water expands when it is
a. heated
b. cooled
c. 'a' and 'b'

4. A pump's suction line is lifting water 10 feet. What force makes water flow into pump?
a vacuum
b: atmospheric pressure
c. capillary attraction

d. all of the above

5. *A*ir in a steam system.

a. increases pressure in boiler
b. reduces pressure
c. obstructs heat transfer
d. 'a' and 'c'

6. At the critical pressure 3,2O6 absolute p.s.i., the water and steam have the same specific weight.

a. true
b. false

7. Natural gas is constituted of

a. hydrogen
b. oxygen
c. methane
d. 'a' and 'b'

8. A material cannot burn unless it is transformed to the gas state.

a. true
b. false

9. Combustion.

a. rapid oxidation
b. slow oxidation
c. atomization
d. 'b' and 'c'

10. Combustion of fossil fuels employ the combining of

a. oxygen
b. hydrogen
c. carbon
d. sulfur
e. all of the above

11. All fuel is transferred to the gas state by

a. heat
b. atomization
c. pressure
d. all of the above

12. Incomplete combustion product in flue gas is

a. carbon dioxide
b. carbon monoxide
c. nitrogen
d. all of the above

13. Sulfur dioxide and condensate in flue gas. The dew point forms

a. sulfuric acid
b. sulfurous acid
c. hydrochloric acid
d. all of the above

14. Enthalpy is the total heat content of a fluid.

a. true
b. false

15. A fluid can be a liquid or gas.

a. true
b. false

16. Sensible heat is the temperature measured by a thermometer.

a. true
b. false

17. Latent heat is heat required to change the state of matter to a

a. solid
b. liquid
c. gas
d. all of the above

18. Impurities in liquids are found in_____state.

a. solution
b. suspension
c. absorbed
d. all of the above

19. What is PH?

a. percent of hydrogen ion content
b. a acid indicator slide chart
c. a alkalinity indicator slide chart
d. 'b' and 'c'

20. Absolute temperature (Farienheight) for a fluid.

a. thermometer temperature (sensible)
b. 'a' plus atmospheric temperature
c. thermometer temperature plus 460°F
d. thermometer temperature subtracted by 460°F

21. Condensation takes place due to the loss of latent heat.

a. true
b. false

22. Why will steam condense when its flow is stopped in a pipe line?

a. molecules stop vibrating due to no motion
b. pressure drop because flow stopped
c. heat transfer takes place without supplemented heat to replenish loss

23. A 'foot pound' is moving one pound weight a distance of one foot.

a. true
b. false

24. Horsepower is the lifting of 33,000 ft. lbs. in one minute.

a. true
b. false

25. Specific heat is the heat required to raise any one pound substance one foot.

a. true
b. false

26. An eleven pound weight suspended three feet has

a. zero potential energy
b. 33 ft. lbs. potential energy
c. 33 ft. lbs. kinetic energy
d. 33,000 ft. lbs. potential energy

27. Condensation of gas will always result when latent heat of evaporation, plus added enthalpy has been removed

a. true
b. false

28. Water is known as the universal solvent and will dissolve anything in time. Water will not wear out or break-down.

a. true
b. false

29. All liquids cannot be compressed if dissolved gasses are not present. Alcohol is an example.

a. true
b. false

***30. P*lace 200°F water into a vented closed vessel, then apply this liquid's surface with a vacuum condition**

a. water will boil
b. dissolved gasses will release from liquid
c. equilibrium will be secured with 'a' and 'b'
d. all of the above

31. The highest temperature of steam will be at_____pressure within a closed vessel.

a. 100 p.s.i.g.
b. 100 p.s.i.a
c. 30" HG
d. 3 bars

32. Heat exchanger such as closed feedwater heaters follow the law of thermodynamics which state; "Heat flows from a cooler body to a warmer body by radiation, conduction and convection." This is why 'counterflow' heat exchangers are used.

a. true
b. false

33. It takes less heat energy to make steam in a vacuum than it will to generate under a 24" HG nitrogen gas layer acting on the water's surface.

a. true
b. false

34. Moisture content in a fuel is desirable as hydrogen and oxygen will be available to increase combustion efficiency.

a. true
b. false

35. Cutting a piece of steel stock with a hacksaw generates heat due to atoms temporarily being split and re-united.

a. true
b. false

36. Viscosity of a liquid decreases with

a. pressure
b. vacuum
c. heat rise
d. all of the above

37. Wet bulb thermometer.

a. liquid filled
b. Fahrenheit or centigrade type
c. ordinary thermometer with wet gauze on bulb
d. all of the above

38. A gas can be acid or alkaline.

a. true
b. false

39. 'Head' is a term to identify energy in a fluid.

a. true
b. false

40. It is impossible for heat to flow from a low to a high temperature. This is the law of thermodynamics.

a. true
b. false

41. A gas compressed with no heat gain or loss is called

a. isothermal compression
b. adiabatic compression
c. thermodynamic compression
d. question is invalid

42. A gas being increasingly compressed at a constant temperature is called

a. isothermal compression
b. adiabatic compression
c. high compression

43. To convert degrees Fahrenheit to degrees centigrade use

a. F°=9/5 (+32°) = C°
b. F =5/9 (C°-3°) = C°
c. C =5/9 (F°-32) = C°

44. 1" of mercury will read to_____ feet of water at 68°F.

a. 1.131
b. 2
c. 4
d. cannot be converted

45. Exhaust steam is dry as long as it hasn't been expanded to a low pressure to such great extent to start condensing. In this state we can still induce steam to a induction turbine.

a. true
b. false

46. A fuel with a high_____content will produce the highest $C0_2$ reading upon combustion.

a. nitrogen
b. hydrogen
c. carbon

47. 15 p.s.i. steam will be_____upon leaving a reducing valve although pressure previously was 230 p.s.i. in a dry saturated state.

a. superheated
b. dry saturated
c. wet saturated
d. 'b' though cooler

48. The highest theoretical CO_2 reading for combustion.

a. 18%
b. 21%
c. 70%
d. 100%

49. Steam at 100 p.s.i. at 383°F will be

a. condensate
b. saturated
c. superheated
d. starting to condense

50. Exhaust steam can be at any pressure. That is why it can be related to being in good quality, but in all reality it is always a lower quality than initial steam. If initial steam to prime mover is dry saturated, steam will be very low quality.

a. true
b. false

51. All steam has moisture content, even superheated steam. A low quality steam can be any pressure or temperature depending on the intended application of its use. Generally speaking, any steam with moisture content higher than dry saturated is rated low quality.

a. true
b. false

Chapter 17

Refrigeration Unlimited

1. The state of a refrigerant beyond the expansion valve is a

a. liquid
b. gas
c. solid

2. State of a refrigerant before the expansion valve in the condenser coil.

a. liquid
b. gas
c. solid

3. Lithium bromide is used in

a. centrifugal compressors
b. ion exchangers
c. absorbers
d. condensate return lines

4. Dichlorodifluoromethane CCl_2F_2. Trade name 'Freon' is used in

a. centrifugal compressors
b. ion exchangers
c. absorbers
d. condensate return lines

5. Centrifugal chillers have (in chill water applications)

a. expansion valves
b. intercoolers
c. condenser and evaporator

6. An absorber would be classified as

a. compressor
b. air conditioner
c. chiller

7. Absorption system requires_____to operate.

a. heat
b. high steam pressures
c. low internal pressure
d. relative humidity over 90%

e. 'a' and 'c'

8. Evaporation coolers require_____ to operate

a. water
b. air flow
c. filters and pump
d. all of the above

9. High relative humidity will render evaporation coolers useless.

a. true
b. false

10. Air flow in one direction delivers cooled air. Air flow reversed will deliver heated air. This describes a

a. centrifugal chiller
b. heat pump
c. absorber
d. evaporation cooler

11. Reciprocating type systems require

a. compressor
b. evaporator and condenser
c. expansion valve and receiver
d. all of the above

12. To purge a air conditioning system means to remove

a. heat
b. non-condensable gas
c. refrigerant
d. all of the above

13. The most toxic refrigerant

a. ammonia
b. CO_2
c. Freon

14. Highest superheating of the refrigerant takes place in the

a. compressor
b. evaporator
c. condenser
d. none of the above

15. One ton of refrigeration capacity (American Standard)

a. latent heat required to make one ton of ice at 32° F
b. 200 BTU/minute.
c. 288,006 BTU/day
d. all of the above

16. Copper pipe may not be used on ammonia systems as it will corrode.

a. true
b. false

17. Some refrigeration systems use a turbine to expand the refrigerant and extract energy instead of an expansion valve.

a. true
b. false

18. Refrigerant gas used in industry.

a. ammonia or Freon
b. carbon dioxide or propane
c. sulfur dioxide or butane
d. all of the above

19. Chromate chemical

a. poisonous
b. prevents corrosion
c. refrigerant
d. 'a' and 'b'

20. Lithium bromide used in absorbers.

a. inflammable
b. prevents corrosion
c. absorbs water vapor
d. all of the above

21. Before starting absorbers and centrifugal refrigeration machines check

a. for vacuum
b. chill (cooler) water flow
c. condenser water flow
d. all of the above

22. The refrigerant in a absorber

a. chromate
b. lithium bromide
c. H_20
d. 'a' and 'b'

23. Lithium bromide in absorbers may crystallize or solidify if

a. air in system
b. generator too hot
c. lithium concentration excessive
d. 'a' and 'c'

24. The absorber's generators' main function is to

a. vaporize lithium bromide
b. vaporize H_20
c. cool lithium bromide
d. cool H_20

Chapter 18

Mechanics Unlimited

1. Test a staybolt on a cold empty boiler to determine if broke.

a. shine a light in telltale hole and observe opposite end
b. tap with a hammer
c. remove the bolts and visually check
d. all of the above

2. You would pipe a natural gas line to a boilers burner

a. close to the floor
b. near the ceiling
c. under the warm boiler setting
d. 'a' or 'b'

3. Natural gas lines to a boiler's burner should be

a. threaded
b. welded
c. flanged
d. compression fitted

4. A natural gas leak will concentrate near

a. ceiling
b. floor
c. midway between 'a' and 'b'
d. six feet from floor

5. To cut male pipe threads use a

a. tap
b. taper die
c. reamer
d. 'flat machine thread' die

6. To remove a thin-wall threaded pipe use a

a. pipe wrench
b. monkey or spud wrench
c. fabric strap wrench
d. hammer and chisel

7. A female pipe thread is made by a

a. tap
b. die
c. 'a' and 'b'
d. reamer

8. After cutting a pipe with a two-wheel cutter, before cutting threads

a. sandpaper the outside diameter

b. taper the outside diameter with a file
c. ream the inside diameter

9. Soldering a copper joint is accomplished by
a. heat
b. capillary attraction
c. cleaned surfaces and flux
d. all of the above

10. A gate valve can be installed in any position or direction to fluid flow.
a. true
b. false

11. Pipe size is measured by the
a. inside diameter
b. outside diameter
c. length
d. thread pitch

12. The nominal diameter of a pipe.
a. inside diameter
b. outside diameter
c. thread length and pitch
d. length

13. Pipe is classified as to strength.
a. standard
b. extra strong
a. double extra strong
d. all of the above
e. 'a' and 'b'

14. Any SAE oil with the manufactures recommended viscosity index can be used in any machine such as a steam turbine.
a. true
b. false

15. Two gears in mesh. The larger is the
a. pinion
b. idler
c. gear

16. Two gears in mesh. The smaller is the
a. pinion
b. idler
c. gear

17. Helical gears are_____than spur gears.
a. stronger
b. quieter
c. weaker
d. 'a' and 'b'

18. To check gear contact wear points, use a
a. feeler gauge
b. dial indicator
c. machinist dye
d. all of the above

19. Before removing an electric motor.
a. open circuit
b. lock and tag panel
c. close circuit
d. 'a' and 'b'

20. Fuses located indoors should be changed often because
a. they deteriorate
b. time delay limit may increase
c. current capacity could increase
d. all of the above
e. none of the above

21. A fuse with a single gap burn-out of 1/4" length has probably been activated by a_____situation.
a. overload
b. short-circuit
c. low level ground fault

22. A duel element fuse has a fuse link for short circuit protection and a thermal element for overload protection.
a. true
b. false

23. If a 18,000 volt circuit breaker opens, reset to see if it trips again then find the cause of circuit opening.
a. true
b. false

24. When starting electric motors, the current flow will actually overload the motors normal running rated amperage on name plate.
a. true
b. false

25. Motor starting consistently blows the fuse.
a. time delay fuse improperly sized
b. grounded condition in motor
c. fuse installed with no time delay
d. any of the above

26. Motor overloading on a routine startup can be expected to reach up to six times normal amperage and not considered unusual.
a. true
b. false

27. A ground fault is a short-circuit and can be difficult to detect because current flow
a. might not increase from normal
b. could appear as a natural overload current
c. current will not flow
d. 'a' or 'b'

28. The piping to a compressor, turbine, engine, etc., should help support the prime mover.
a. true
b. false

29. Helical gears are left handed and right handed as determined by sight along the axis.
a. true
b. false

30. To prevent piston rod packing from wearing out prematurely
a. install grease fittings
b. let some fluid leak by to cool packing
c. tighten stuffing box until leak stops
d. reduce the number of packing rings by 50%
a. use a smaller size packing

31. A mechanical shaft seal leaks.
a. replace seal with a new one
b. adjust the packing nuts
c. grease the carbon ring
d. heat seal with a torch

32. To grease a non-sealed roller bearing.
a. always grease with shaft rotating
b. always grease with shaft stationary
c. 'a' or 'b'
d. remove bearing then pack with grease

33. Roller bearing overheating on a variable drive motor.
a. shut down machine
b. grease the bearing
c. spray water on bearing casing
d. slow motor to low speed

e. 'd' - 'b' - 'a' in this order

34. Two machines connected to a flexible coupling
a. need not to be aligned
b. must be aligned
c. coupling must rotate clockwise
d. 'a' and 'c'

35. An alloy metal.
a. molybdenum
b. stainless steel
c. tungsten
d. all of the above

36. Case hardened steel shaft will have a hard exterior and a soft interior.
a. true
b. false

37. The foundation must be thicker and stronger for a steam or diesel engine than a steam turbine of the same size and weight due to reciprocating forces.
a. true
b. false

38. Battery charging produces explosive_____gas.
a. nitrogen
b. oxygen
c. sulfuric oxide
d. hydrogen

39. When installing a gasket in a pipe flange, it is best to only tighten the flange enough to stop leakage rather than tighten nuts to full torque rating of bolt stretch.
a. true
b. false

40. You are working alone. A feed pump electric motor catches fire. It is best to
a. start standby pump
b. shut down boiler fires
c. start an injector pump
d. extinguish fire with emergency water fire pump

41. A valve for fine control.
a. globe
b. needle
c. pet cock
d. plug cock

42. To extinguish an electrical panel fire, use
a. water
b. foam
c. carbon dioxide
d. baking soda

43. SAE represents
a.. Society of Automotive Engineers
b. Society of Automotive Engines
c. Society of Automotive Emollients
d. Society of Automatic Engines

44. A pitot tube is used to measure
a. specific gravity of battery acid
b. flue gas temperature
c. brine concentration
d. air velocity

45. A pigtail to a steam pressure gauge is to allow for pipe expansion and contraction.
a. true
b. false

46. 'Saybolt Universal' charts refer to
a. fuel BTU value
b. flow characteristics
c. moisture content
d. torque strength

47. A three wheel type pipe cutter will not cut steel pipe unless rotated completely in same direction due to wheel cutting edge angle.
a. true
b. false

48. A thin light oil is used on low speed heavily loaded bearings.
a. true
b. false

49. Instrument used to measure specific gravity of a liquid.
a. viscosimeter
b. manometer
c. hydrometer
d. calorimeter

50. To remove a frozen bearing on a round shaft, first
a. cool the bearing
b. heat the shaft
c. heat the bearing
d. 'b' and 'c'

Chapter 19

DIESEL ENGINES

DIESEL INTRODUCTION

Diesel engines are made in many sizes and horsepower ratings. From 30 horsepower to as high as 40,000 and higher. They can run on distillate or gaseous fuels and many can be switched to one or the other (or both) depending on fuel availability and cost. Applications range from electric generation to ship propulsion though frequently used for emergency electrical generation back-up in hospitals, sky risers and industry. Stationary, marine, and locomotive engines require maintenance and observation by trained personnel. This study guide will deal with Stationary diesel plants in particular, but may easily be related to any category.

Like steam turbines driving generators, diesel units are rated in kilowatts (746 watts = 1 horsepower).

There are two basic classes of engines; 2 stroke and 4 stroke. Both are 4 cycle engines. Cycle pertains to cycle of events such as intake, compression, power and exhaust. A 2 stroke engine accomplishes all four cycle of events in 2 strokes of the piston, or one revolution of the crankshaft. A four stroke engine accomplishes all four cycles, but requires the piston to stroke four times, or two crankshaft revolutions.

Supercharging refers to pressurizing air into

the cylinders above atmospheric pressure. One method used is employing a air compressor - called a 'blower' - driven by any source other than exhaust gas such as an electric motor, or mechanically driven off of the engines crankshaft. Another method is called 'turbo-charging' which utilizes a turbine in the exhaust to drive a blower. Note that both have blowers but are driven by different means.

I recommend for the student to purchase Diesel repair manuals or write to Diesel manufactures for literature. You can find them in the public library listed in the Thomas Register. Find a Diesel plant and visit often. If you live near a shipping port or locomotive railway, talk with Engineers. This test will cover many types of Diesels, so learning about Diesel tractor trailer engines only will not qualify you for the Diesel engineers license, but the test will guide you as to what to learn and expect on a exam.

I also recommend studying the 'Auxiliary' and 'Physics' sections in this book including 'Turbines' and 'Steam Engines'. You should study such related equipment such as pumps, heat exchangers, governors, turning gear, safety valves and air compressors. A knowledge of steam engineering will be very helpful, if not a necessity.

In the test, you will come across a few terms:
Bar - 14.5 p.s.i.
T.D.C. - Top Dead Center
B.D.C. - Bottom Dead Center
B.T.D.C. - Before Top Dead Center
A.T.D.C. - After Top Dead Center
D.C. - Dead Center- extreme end of piston stroke, crankpin in line with connecting rod on zero angle. No power can be applied to produce circular motion. Detonation or combustion timing on this angle will damage bearings or break crankshaft.
D.C. may also refer to direct current electricity.

In the test all questions will refer to a standard compression ignition oil fired diesel engine unless specified in the question 'gas' diesel, and this will not be gasoline but gaseous fuel such as natural gas and may be pilot oil or spark ignition fired.

STARTING A DIESEL

CHECK:

1. Fuel level, supply, drain filters and prime system.
2. Fill engine jackets with water, expansion tank 3/4 full.
3. Vent cooling water pumps and start pumps. Close vents when primed.
4. Drain scavenging air box of condensate and blowers.
5. Oil level and prime until oil appears at rocker levers. Fill governor.
6. Pour oil over rockers and valve mechanisms and fill air maze loop.
7. Fuel supply free of condensate.
8. Exhaust line open and drained.
9. Cylinder test valve or vent open.
10. Engine lever in injector stop position.
11. Turn engine with turning gear. No fuel or H_2O at cylinder vents. Check for binding and disengage turning gear.
12. Close cylinder test valves.
13. Place operating lever to injector full with latch connected to injector linkage.
14. Open air starting valve. 10 second should start, then throttle speed to idle.
15. Check oil pressure and warm up slowly raising Rpm's.
16. Disengage engine operating lever from injector linkage and latch into governor run position.
17. When up to speed, close air starting valves and bleed lines.
18. Add load slowly.

TO STOP ENGINE

1. Reduce load slowly.
2. Run under governor control.
3. Engage operating lever to injector lever

and move to injector stop position slowly (bring speed down gradually).

4. Open drains when engine stops.
5. Rotate engine on turning gear.

Read the manufacturer instructions on how to start, stop, operate and maintain the engine.

Once a diesel engine is setup it will normally run trouble-free for many years only requiring routine oil changes, injector timing, fuel, oil and air filter changes.

The greatest danger is overspeeding, overloading and running out of lubricating oil. These engines will explode unmercifully throwing metal fragments like a bomb. Keep crankcase vents open and clear to prevent fuel vapor accumulating in the crankcase as this too can cause the case to burst. However, most large diesels have explosion doors or 'weak points' such as; inspection plates that will burst. Nevertheless, any burst is dangerous. Never be complacent when running a large diesel. Always be aware of rotating elements and never take any safety guards off an engine without tagging and locking-out the starting control panel.

WRITTEN EXAM TIPS

Never be the first to finish the test. Skip the questions you are not sure of and continue the exam, then later go back to those you couldn't answer. Remember to read the question carefully. The correct answer is commonly always within the question. Look for the 'key' word which may alter the meaning of the question. Example: In question # 2 in Diesel exam all answers are correct until you see the word 'in-line.' Only one answer can now be correct and that is 'b'. After finishing the exam, review your answers, but do not make any changes unless you are absolutely certain you marked the wrong answer. You're first choice may be the correct choice. Read the question carefully and 'think hard'. Always assume each question is a trick question because they usually are.

1. Two main types of diesel engines.
a. 2 stroke and 4 stroke
b. 2 cycle and 4 cycle
c. fuel oil and gasoline

2. A vertical in-line engine.
a. V-configuration
b. cylinders one behind the other
c. delta type

3. An opposed piston type engine.
a. 1 piston in 2 cylinders on 2 crank pins
b. 2 cylinders, 2 pistons in line with one crank
c. 2 pistons in 1 cylinder with 2 crankshafts

4. A horizontal engine.
a. 2 cylinders with 1 piston in each, 1 crankshaft located between pistons. Crank is horizontal
b. any engine on a 45° angle
c. circular horizontal mounted cylinders connected to a vertical crankshaft only

5. Delta engine.
a. 3 cylinders, 3 crankshafts, 6 pistons
b. 3 cylinders, 1 crankshaft, 3 pistons
c. 2 in-line V-8 engines connected to same driveshaft

6. A four-stroke cycle engine needs____for control of cycle events.
a. ports
b. valves
c. blowers

7. A two-stroke type engine utilizes____for control of cycle events.
a. blowers
b. valves and cams
c. ports or valves

8. Volumetric efficiency.
a. pressure of intake air divided by exhaust pressure
b. fuel input divided by horsepower output
c. volume of air drawn in per stroke divided by volume swept by piston

9. Intake poppet valves are mostly used in two-stroke, four cycle diesels.
a. true
b. false

10. The exhaust valve runs cooler than intake valve.
a. true
b. false

11. A larger intake valve will
a. boost volumetric efficiency
b. raise compression ratio
c. 'a' and 'b'
d. Reduce engine jacket cooling water temperature

12. Two intake and two exhaust valves in one cylinder will
a. increase low speed power torque
b. require four camshafts to operate valves

c. run cooler and increase air inlet and exhaust outlet flow rates
d. Create combustion mixture swirl pattern

13. A two-element inlet valve is for a
a. high speed engine
b. low speed engine
c. gas fired engine
d. no such valve exists

14. Camshafts are used to open and close
a. air starter valves
b. fuel valve (natural gas)
c. exhaust valve and injector
d. all of the above

15. Camshaft speed in a 4-stroke engine.
a. same speed as crank
b. twice the speed of the crank
c 1/4 crank speed
d. 1/2 crank speed

16. Camshaft speed in a 2-stroke engine.
a. same speed as crank
b. twice the speed of the crank
c. no cam in any 2 stroke engine
d. 1/2 crank speed

17. Some engines have no push rods to operate valves.
a. true
b. false

18. Some camshafts are assembled by bolting them together.
a. true
b. false

19. Helical gears are used because they
a. never wear out
b. are quieter
c. can't chip or crack
d. give warning before failure

20. Base circle on a cam is
a. the total height of the lobe
b. average lift of cam lobe
c. area of no lift

21. Scavenging is
a. cooling cylinders
b. blowing air out exhaust header outlets
c. replacing exhaust with air

22. Blower scavenging takes place in a 4-stroke engine when
a. piston at bottom dead center
b. piston at mid-stroke
c. piston at top dead center

23. Crankcase scavenging in a single cylinder 2-stroke engine is accomplished by
a. blowers
b. turbochargers
c. piston displacement
d. 'a' or 'b'

24. Blower scavenging is accomplished by
a. engine connected drive source
b. lobed impellers
c. 'a' and 'b'
d. positive displacement gears

25. Engines with no valves can be scavenged by the use of
a. blowers or turbos
b. piston displacement
c. ports in cylinder
d. all of the above

26. Valve scavenged engine has
a. intake valve and air starting for reverse rotation
b. exhaust valves and ports
c. cylinder ports only

27. Crossflow scavenging.
a. utilizing two blowers
b. air enters side of cylinder and out other side in one direction
c. uses exhaust gasses to purge cylinders
d. air changes direction in cylinder

28. Loop scavenging.
a. air enters and leaves same side of cylinder after changing direction
b. uses exhaust gasses to purge cylinders
c. two-stage blower system to purge cylinders and supply combustion air

29. Uniflow scavenging.
a. air traveling in one direction
b. air reversing direction in cylinders
c. turbo blower in exhaust header to create vacuum in cylinders

30. Supercharging cylinders.
a. installing high performance pistons
b. burning a rich fuel mixture
c. volumetric efficiency increase by compressing of air

31. A very low speed engine will most likely employ the
a. blower
b. turbocharger
c. 'b' in exhaust and 'a' in intake system

32. High speed engine will employ a
a. piston crankcase scavenging design
b. blower
c. turbocharger

33. Turbocharger is driven by
a. gears or chains
b. belts and pulleys
c. exhaust gasses

34. Engine with open intake and exhaust valves overlapping accomplishes
a. scavenging
b. cylinder cooling
c. 'a' and 'b'
d. valves cannot overlap

35. Blowdown system utilizes_____ to increase turbo output
a. pulsation of exhaust
b. exhaust gas bypassing turbo to atmosphere
c. a steady flow of exhaust gas

36. Turbo that operates at a steady exhaust gas rate is a

a. pulse converter type system
b. blowdown type system
c. constant pressure type system

37. Pulse converter utilizes
a. blowdown effects with a velocity increase
b. constant pressure effects
c. 'a' and 'b'

38. Compressing air
a. reduces its volume
b. raises its temperature
c. 'a' and 'b'
d. increases crankcase oil temperature

39. Intercooling or aftercooling compressed air will
a. increase volumetric efficiency
b. cool cylinders, valves, pistons
c. reduce detonation probabilities
d. all of the above

40. A gas diesel burns
a. natural gas
b. gasoline with no octane
c. oxygen
d. JP-4 aviation fuel

41. A gas diesel is much more perceptual to
a. detonation
b. wipe oil from cylinder walls
c. clog injectors with methane
d. overspeeding

42. A diesel engine will likely have a
a. spark plug
b. carburetor
c. camshaft
d. none of the above

43. Another name for a turboexpander is a
a. turbocooler
b. afterburner
c. turbocompressor
d. intercooler

44. A turboexpander cools air by
a. expanding in a turbine
b. a shell and tube heat exchanger
c. passing air into exhaust system

45. Cooling air within the cylinder is accomplished by (supercharged conventional engine)
a. compression increase by installing 'high dome' pistons
b. decreasing cooling water to engine jackets
c. cutoff air before bottom dead center of piston intake stroke
d. none of the above

46. Some engines are started with a mechanical blower then switched to a turbine once up to operating speed.
a. true
b. false

47. Compressed air for starting can be introduced to
a. turbocharger
b. injectors
c. cylinders
d. 'a' and 'c'

48. When supercharging a non-supercharged engine a larger water cooling system must be installed
a. true
b. false

49. Clearance volume.
a. space between piston dome and cylinder combustion chamber
b. compression ratio
c. total area swept by piston in square inches

50. When supercharging a non-supercharged engine you must change
a. valve and injector timing
b. clearance volume
c. injector volume (increase)
d. all of the above

51. Injectors or injector systems
a. raise fuel pressure
b. meter fuel and timing
c. atomize fuel
d. all of the above

52. Compression ratio can be increased by
a. installing a blower
b. increasing piston stroke
c. installing a turboexpander
d. install undersize piston wrist pin

53. Controlling valve lift and valve timing will
a. govern speed of engine
b. increase or decrease initial cylinder pressure on power stroke
c. 'a' and 'b'
d. not feasible in diesel engines

54. Injector cutoff is controlled by the
a. camshaft
b. governor
c. lubricating oil supply pump
d. fan belts

55. Cutoff is
a. shutting off the engine
b. changing over fuel pumps
c. interrupting fuel or air admission to cylinders
d. opening turbocharger wastegate valve

56. An eccentric is used to
a. adjust piston stroke
b. run the governor
c. operate and change timing events
d. trigger overspeed trip device

57. Injectors are_____atomizers.
a. air
b. gas
c. hydraulic
d. pneumatic

58. Cutoff within an injector (relieving back pressure)
a. allows check valve to close quickly
b. allows check valve to open quickly

c. no check valve in injectors

59. A dirty or clogged injector
a. willcause a rich mixture and produce smoky exhaust
b. may not inject fuel at all
c. can overheat and melt injector
d. all of the above

60. Fuel is kept flowing in injectors to
a. lubricate spray tip
b. accomplish cooling
c. warm fuel in supply tank
d. 'b' and 'c'

61. Turbulence inside diesel cylinders is desirable for combustion.
a. true
b. false

62. Compression temperature corresponds with air pressure.
a. true
b. false

63. Injector that pumps, meters, and times the fuel charge is a
a. uniflow injector
b. unit-injector
c. safety injector

64. Fuel pumps are of the_____type.
a. centrifugal
b. rotary or reciprocating
c. vane
d. turbine

65. Dual fuel engine burns
a. natural gas
b. fuel oil
c. 'a' and 'b' at the same time
d. 'a' and 'b' normally not at the same time, not including pilot oil injection

66. Engine cylinder condition that needs a fine atomized fuel spray.
a. cylinder with little turbulence
b. cylinder with heavy turbulence
c. cylinder with worn oil control ring

67. One crankshaft revolution consists of _____degrees.
a. 180
b. 90
c. 360
d. 120

68. Cylinder oil control rings are located on
a. piston
b. cylinder
c. transfer ports

69. Cylinder transfer ports are for
a. oil control
b. scavenging
c. water jacket connections

70. Smoky exhaust indicates
a. loss of compression
b. late injector timing
c. dirty injector spray tip
d. all of the above

71. A small charge of ignited fuel used to ignite main cylinder charge is a
a. open type combustion chamber engine
b. precombustion chamber type engine
c. rotary type engine

72. Mean effective pressure.
a. average pressure acting on piston on all strokes
b. average pressure in ft.lbs. on crankshaft
c. average pressure acting on piston power stroke
d. compression pressure

73. Cycle of events refer to_____ strokes.
a. intake
b. compression
c. power
d. exhaust, 'a' and 'b' and 'c'

74. Crankshaft makes one revolution is called a
a. throw
b. stroke
c. 180° throw

75. Four cycle of events take place in two crank revolutions is a
a. four stroke engine
b. two stroke engine
c. rotary engine

76. A stroke is_____of a crankshaft revolution.
a. complete rotation
b. one half rotation
c. two thirds rotation
d. 90°

77. Piston travels from top to bottom in a cylinder is a
a. stroke
b. throw
c. complete crankshaft revolution

78. Four cycle of events take place in one crankshaft revolution is a
a. four stroke engine
b. two stroke engine
c. rotary engine
d. steam or diesel engine

79. Medium in which piston travels.
a. crankcase
b. cylinder head
c. cylinder sleeve
d. all of the above

80. Pistons receive initial energy then transmits through_____to crankshaft.
a. wrist pin
b. connecting rod
c. 'a' and 'b'

81. Camshafts are driven by
a. an external source
b. gears or chains
c. 'a' and 'b'
d. serpentine belts

82. A natural gas engine most likely has a
a. ignition spark or pilot oil
b. carburetor
c. none of the above

83. Diesel engine intake stroke is filled with
a. air and fuel
b. fuel only
c. air only

84. Preignition is readily noticeable by
a. low oil pressure
b. knocking sound
c. late ignition timing
d. cooling water temperature rise

85. Detonation is a term for
a. incomplete combustion
b. preignition
c. post ignition

86. If cylinder safety valve lifts often, it could be caused by
a. preignition
b. detonation
c. high oil pressure and post ignition
d. 'a' and 'b'

87. Preignition is caused by
a. injector timing too early
b. inferior fuel
c. 'a' and 'b'
d. blower failure

88. Detonation is a result of
a. preignition
b. bent connecting rod
c. worn piston skirt
d. all of the above

89. Relief valve on a fuel oil supply is to
a. protect oil pump from bursting
b. maintain a set design fuel supply pressure
c. circulate fuel oil
d. all of the above

90. Drains are located on a large diesel to prevent
a. hydraulic damage from oil or water
b. overspeeding engine when starting
c. no drains on a diesel engine

91. Drains are located on
a. blower
b. exhaust
c. cylinders
d. any or all of the above
e. none of the above

92. It is desirable to have high compression ratios in a diesel engine.
a. true
b. false

93. Compression ratio.
a. area a piston crown multiplied by combustion chamber area at end of stroke
b. piston size in square inches multiplied by cylinder head size in square inches
c. volume of cylinder at beginning of stroke added to volume at end of stroke

94. Natural gas engines are usually ultra high compression engines.
a. true
b. false

95. Inferior fuels are used in
a. high compression engines with low turbulence
b. low compression engines with high turbulence
c . high speed, 'a' only
d. V-8 type engines only

96. Aspiration means in any internal combustion engine.
a. exhausting to atmosphere
b. breathing of intake stroke
c. 'a' and 'b' by piston displacement

97. Diesels are external combustion engines.
a. true
b. false

98. Normally aspirated engine means
a. piston created vacuum creating a natural air flow
b. piston created pressure causes a natural air flow
c. 'a' and 'b'
d. none of the above

99. An externally fired engine.
a. diesel
b. steam
c. gasoline
d. natural gas

100. Pressure gauges are located on a fuel oil
a. suction line
b. discharge line
c. oil return line from relief valve
d. all of the above

101. Fuel oil supply tanks must be
a. above engine
b. below engine
c. vented to atmosphere
d. 'a' and 'c'

102. Cooling water pumps to engine jackets are of the_____type.
a. centrifugal
b. reciprocating single acting
c. rotary
d. gear

103. The governor controls engine
a. speed
b. exhaust valves
c. oil pressure
d. all of the above

104. A diesel with underground fuel tank has a minimum of_____fuel pump(s).
a. one
b. two
c. three

105. Air intake filter should be located
a. outside near exhaust outlet
b. outside with suitable protection from elements
c. always near operating floor level

106. By installing a blower, compression ratio increases.
a. true
b. false

107. Accessories on an exhaust line permit
a. drain line for H_20 and expansion joint
b. heat exchanger
c. silencer
d. all of the above
a. no accessory can be installed

108. Starting a diesel may employ
a. compressed air
b. battery powered motor
c. hand crank or jacking bar
d. 'a' and 'b'

109. Oil filter automatic safety device.
a. pressure gauge
b. bypass valve
c. 'a' and 'b'
d. none of the above

110. A low compression gas diesel operates on
a. near perfect fuel to air ratio
b. very rich mixture

c. lean mixture

111. A high compression diesel operates on

a. near perfect fuel to air ratio
b. very rich mixture
c. lean mixture

112. Pilot oil is used for

a. gas engines
b. steam or 'hit and miss' engines
c. non-blown fuel oil diesels only
d. 'a' and 'c'

113. When does valve overlap take place?

a. both intake and exhaust valves open together on power stroke
b. when intake and exhaust cam lobes are on base circle on any stroke
c. intake valve opening, exhaust valve closing at beginning of intake stroke

114. Why is valve overlap used on diesel engines?

a. scavenges combustion chamber
b. helps cool exhaust valves and pistons
c. increase power output
d. all of the above

115. What is a crank pin?

a. journal bearing surface
b. webs that hold the journal
c. another name for wrist pin

116. Non-supercharged scavenging takes place in a four stroke, four cycle engine.

a. piston on top dead center at beginning of intake stroke
b. piston on bottom dead center at beginning of exhaust stroke
c. piston at mid-travel on exhaust stroke

117. An inferior fuel will not withstand compression and will ignite before it should, burns slowly, carbonizes combustion chambers, difficult to atomize and has a low BTU per unit measure.

a. true
b. False
c. 'a' but not difficult to atomize fuel

118. Cooling water engine jacket expansion tank should be connected

a. . at the lowest point in the system
b. on the outlet line to engine
c. on the inlet line to engine
d. 'a' and 'c'
e. 'b' and 'c' for expansion and venting of lines

119. A high grade fuel burns clean, rapid, withstands compression, easily atomized, has a high BTU per unit measure.

a. true
b. false

120. A transfer fuel oil pump

a. supplies oil to injector
b. delivers oil to injector fuel oil pump or day tank
c. circulates oil to main bearings

121. What does pilot oil do?

a. prime the injectors
b. start ignition
c. lubricate injectors and camshaft lobes
d. 'c' and lube crankshaft journals

122. Majority of heat leaves the intake and exhaust valves when

a. valve leaves its seat (opening)
b. valve contacts seat (closed)
c. valve is exposed to exhaust gas

123. A intake and exhaust valve should periodically rotate during operation to

a. prevent an out of round valve seat
b. help retain a trace of oil in valve guide
c. 'a' and 'b'
d. valves do not rotate

124. Why do diesels run at a lean air / fuel ratio?

a. to insure complete combustion
b. to cool pistons only
c. so blower will not over pressurize
d. to prevent explosion in exhaust line

125. What will cause detonation preignition?

a. inferior fuel
b. injector timing too early
c. carbon build up in combustion chamber
d. blown head gasket
e. 'a' and 'b' and 'c'

126. If valves and seats prematurely wear often

a. install a 'stellite' replacement valve seat
b. check valve lash
c. check cam lobe wear
d. any or all of the above

127. If you see steam condensing from exhaust, what is wrong?

a. H_20 entering scavenging air
b. crack in cylinder cooling jacket
c. H_20 contaminated fuel
d. blower intercooler tube leak
e. all of the above

128. What will cause unit injector failure?

a. low fuel oil pressure
b. overheating or tight plunger clearance
c. dirty fuel oil or corrosion of nozzle
d. all of the above

129. Engine condition monitors can be found on some large diesels on

a. cylinder walls
b. pistons
c. exhaust system
d. all of the above

130. Piston crowns can be cooled by

a. oil
b. water
c. 'a' or 'b'

131. Where will you find a condensate separator on a typical diesel engine?

a. beyond the air cooler from blower
b. before the air cooler from blower
c. upper oil level in crankcase

132. The crosshead allows reciprocating motion to be converted to rotary motion via the crankshaft.

a. true
b. false

133. An engine run below rated speed will induce higher cylinder pressures, increased loading on the bearings, and poor fuel economy.

a. true
b. false

134. Some slow speed engines can run as low as 90 RPM.

a. true
b. false

135. If one of three turbochargers were to breakdown, what would you do?

a. run at half-speed
b. stop engine and repair
c. increase engine speed to compensate

136. What would make one cylinder stop firing?

a. blown head gasket, no valve clearance
b. injector failure, cracked piston crown
c. worn piston rings or valve seats
d. main fuel oil pump failure
e. 'a' and 'b'and 'c'

137. Another name for a wrist pin.

a. crosshead guide
b. gudgeon pin
c. journal
d. web

138. A engine with a rotating piston will

a. distribute oil uniformly
b. reduce oil consumption
c. reduce stress and wear
d. pistons can't rotate in any engine
e. 'a' and 'b' and 'c'

139. Where will the oil enter to lube a wrist pin and piston crown?

a. splash lubricated from crank contact with oil bath
b. from crank, connecting rod drilled oil passage or auxiliary line
c. from cylinder walls
d. all of the above

140. What is the 'big end' bearing?

a. outer main bearings
b. inner main bearings
c. lower connecting rod bearing
d. 'a' or 'b'

141. You can easily run a high speed diesel engine on heavy fuel oil.

a. true
b. false

142. Crankshaft suddenly broke in half, but bearings not worn.

a. dirty oil and injector failure
b. cylinder pressure too low and timing too late
c. main bearings out of align due to foundation deflection

143. Another name for a ground level fuel tank is a 'day tank'.

a. true
b. false

144. Heavy fuel oil must be_____ before using in diesel engines.

a. purified
b. heated
c. clarified
d. all of the above

145. Heat from the exhaust is used to generate steam to heat fuel oil via heat exchangers in exhaust system.

a. true
b. false

146. Always warm-up a diesel before applying load.

a. true
b. false

147. Why would you find a auxiliary oil fired steam boiler near a heavy fuel oil fired diesel?

a. to raise lube oil temperature
b. to thin oil on start-up via steam
c. to inject steam into cooling water jackets
d. to start engine by injecting steam in cylinders
e. all of the above

148. Crosshead engines prevent

a. contamination of crankcase oil from combustion products
b. contamination of crankcase oil from piston cooling water
c. 'a' and 'b'

149. A piston scavenged, and supercharged type engine, will run even when superchargers fail.

a. true
b. false

150. The thrust bearing is located on the_____and lines-up___clearance.

a. crankshaft - horizontal
b. crankshaft - axial
c. 'a' and 'b'

151. Bore cooling refers to drilled passages.

a. true
b. false

152. Bore cooling results in an increased safety margin with regard to thermal loading and stress.

a. true
b. false

153. A nonreversible engine must be used when generating electricity.

a. true
b. false

154. A diesel engine burning heavy high viscosity fuel will have water cooled fuel nozzles.

a. true
b. false

155. If injector timing is too far out of adjustment it will be necessary to adjust camshaft or lobe position.

a. true
b. false

156. The 'small end' bearing.
a. journal on crankshaft
b. thrust bearing on crankshaft
c. main bearing on crankshaft
d. wrist pin bearing

157. If a bearing started to overheat it is best not to stop the engine, but to rotate it slowly so it will cool and avoid freezing on the shaft if oil is available.
a. true
b. false

158. Another term for the fuel oil return line from injector.
a. jumper tube
b. spill pipe
c. drain
d. 'a' or 'b'

159. Some diesel cam lobes can be relocated on the camshaft for timing with a hydraulic tool.
a. true
b. false

160. How can you tell at a glance an engine may most likely be reversible?
a. auxiliaries are connected to engine, oil pumps, etc.
b. engine has duel fuel capacity
c. auxiliaries are not connected to engine, oil pumps, etc.

161. Auxiliaries are usually driven at the crankshaft
a. thrust bearing end
b. vibration damper end
c. turning gear

162. Injector can fail if fuel strainer (jet filter) is plugged up.
a. true
b. false

163. Loss of turbocharger efficiency, though turbocharger is in good mechanical order. What is wrong?
a. bearing failure
b. drain open
c. dirty turbine blades
d. 'b' or 'c'

164. Loss of lubricating oil pressure, but oil pump is working okay.
a. suction line leak
b. relief valve stuck open
c. strainer on 'a' is fouled
d. any of the above

165. Where would you find an expansion tank?
a. on fuel oil system
b. on jacket cooling water system
c. just beyond the blower air box
d. all of the above

166. What is a air maze?
a. antisiphon loop for lube oil
b. antisiphon loop for fuel oil
c. antisiphon loop for cooling water
d. filter for blower air

167. Where do you attach a compression gauge?
a. on cylinder wall
b. cylinder test valve at head
c. in exhaust valve port
d. all of the above

168. Vernier adjustment on the engine operating lever is for
a. fine adjustment of speed under manual control
b. fine adjustment of speed under automatic control
c. to shutdown injectors in emergency

169. A vibration damper is located on the
a. camshaft
b. crankshaft
c. crosshead
d. foundation and 'c'

170. A typical engine operating lever control will have
a. governor 'run' and 'stop'
b. injector 'stop' and 'full'
c. linkage pawl to 'a' and 'b'
d. all of the above

171. Binding in the injector control racks. What is wrong?
a. injectors not seating properly in cylinder head or too tight
b. micro adjustment rod out of alignment, bent, or too tight
c. injector rack teeth worn or chipped
d. all of the above

172. Where will you find timing marks on a diesel engine?
a. flywheel
b. cam gears
c. blower gears
d. all of the above

173. Journal is the contact surface on a shaft which bearings ride.
a. true
b. false

174. A pyrometer is used to check the temperature of each cylinders' exhaust.
a. true
b. false

175. The difference between any two cylinders exhaust should not exceed 50°F.
a. true
b. false

176. To adjust any individual cylinder exhaust temperature, adjust the
a. governor needle valve
b. injector micro adjustment screw
c. timing chain or gear clearance
d. all of the above

177. The air box should be drained
a. only when engine is stopped
b. only when engine is at full load
c. only when engine is idling
d. any of the above and as often as necessary

178. Excessive lube oil pressure will cause
a. vibration
b. bearing damage
c. 'a' and 'b'
d. no damage and is desirable

179. Stop a diesel in an emergency.
a. press shutdown button on governor
b. move operating lever to governor 'stop' position
c. shutdown fuel pumps
d. any of the above

180. Engine indicator card measures
a. horsepower output
b. average compression ratio pressure
c. cylinder expansion temperatures

181. A diesel with forced pressurized air to cylinders will require
a. less fuel
b. more fuel
c. larger cooling system
d. 'b' and 'c'

182. What will cause engine to backfire?
a. low fuel pressure
b. injector spindle sticking
c. hole in piston crown

183. Before starting a diesel generator set
a. main breaker should be closed
b. check generator voltage
c. open switch gear breaker
d. check blower pressures

184. To adjust cooling water temperature.
a. slow down engine
b. reduce exhaust temperature
c. reduce injector pressure
d. start cooling water pump

185. What causes high oil temperature?
a. combustion temperature too hot
b. high blower pressures
c. engine load
d. cylinder relief valve closed

186. If oil were in the air box it could cause diesel to_____when starting (2-stroke engine)
a. reverse rotation
b. backfire into blower
c. backfire into exhaust
d. stop

187. Engine rotating on turning gear. Make sure_____is operating.
a. oil pump
b. water pump
c. governor
d. fuel relief bypass

188. How would you begin to tighten a 12 bolt flange? (Numbers represent a clock dial).
a. 12 - 6 - 3 - 9
b. 12 - 3 - 6 - 9
c. 12 - 3 - 9 - 6

189. Cooling a diesel. Use a_____for engine jacket water.
a. induced draft cooling tower
b. forced draft cooling tower
c. spray pond
d. any of the above

190. To test a diesel overspeed trip.
a. increase speed
b. push governor stop
c. shut off oil to governor
d. decrease load

191. If turning gear was engaged and found that it has turned diesel in opposite rotation, what will happen?
a. pistons will bend valves
b. bearings will scrape
c. nothing will happen
d. compression pressure will lift cylinder safety valves

192. Lube oil can be reused after
a. separation settling
b. centrifuging
c. gravity filtration
d. heating to precipitate water

193. When placing a diesel's generator on the bus. First check
a. voltage
b. amperage
c. shaft rotation speed

194. What is meant by a diesel 'steam engining'?
a. expansion taking place in cylinder
b. engine running hot exhaust temperature
c. water entering cylinders
d. injector cutting off too soon

195. The exhaust from a diesel can be used to
a. generate hot water for heating
b. cool lube oil at low speed
c. preheat combustion air
d. dewater lube oil

196. Main cooling jacket has a leak in a large diesel. How can you tell?
a. white smoke from breather vent
b. cooling tower makeup increased
c. oil level dropping
d. water in governor oil

197. Another term for the injector rack.
a. spindle
b. plunger
c. link
d. microscrew

198. Prime reason to change lube oil when engine is hot.
a. to speedup drainage
b. allow most oil to drain
c. keep impurities in suspension
d. never drain oil when hot

199. Fuel that does not burn well should be used in
a. low compression engine
b. high compression engine
c. two-stroke, four-cycle engine without blowers

200. Low cylinder compression.
a. tight valve clearance
b. tight injector seating
c. late ignition
d. dirty injector spray tip

201. Before testing overspeed trip on diesel generator set
a. reduce load
b. check oil pressure
c. check bearing temp.
d. open breaker
e. 'a' and 'b' and 'c'

202. Saybolt universal refers to

a. a bolted flexible joint
b. fuel volatility
c. oil viscosity
d. cylinder temperature chart

203. Worn intake valve stem could cause_____, but valve is seating tight just the same.
a. engine vibration
b. preignition
c. backfiring

204. If cylinder compression temperature reached 1,OOO°F what is wrong?
a. too much fuel
b. too much air
c. no oil to cylinders
d. nothing is wrong

205. What would you do if compression temperature reached 1,000° F. ?
a. open emergency water cooling to engine jackets
b. reduce fuel and air supply
c. increase lube oil pressure
d. none of the above
e. 'c' and 'b' and 'a'

206. Which main bearings' clearance should be the closest tolerance on a inline eight-cylinder engine?
a. twocenter bearings
b. outboard bearings only
c. every other bearing to allow for expansion
d. all bearings the same

207. High bearing temperature will raise
a. oil pump pressure
b. oil temperature
c. safety relief valve
d. 'a' and 'b'

208. Adding a shim to connecting rod bearing will
a. decrease compression
b. increase compression
c. decrease bearing clearance

209. You keep open_____(1) when engine is_____(2).
(1)
a. drains
b. cylinder vents
c. oil filter bypass

(2)
a. starting
b. on turning gear
c. at 1/2 speed

210. If rocker arm bearing clearance increased, it would cause unit injector to
a. fire early
b. fire late
c. preignite
d. backfire

211. Eight cylinder diesel refuses to start.
a. compression loss in three cylinders
b. injector rack disengaged
c. no oil in governor
d. injectors stuck halfway closed

212. Water in the oil.
a. shut down engine
b. increase oil temp 30°F
c. add oil so water will overflow out breather
d. decrease speed then perform 'b' or 'c'

213. You pump cooling water from a diesel engine, not through it.
a. true
b. false

214. White smoke from a diesel's exhaust.
a. water entering main bearings
b. water pressure too high
c. burnt exhaust valve
d. injectors wide open

215. High oil and bearing temperature on gear reduction temperature gage. What will you do?
a. disengage gears
b. reduce load
c. cool oil via heat exchanger
d. 'b' and 'c'

216. Piston scoring is caused by
a. load
b. oil
c. heat
d. clearance failure

217. Oil entering the air box will cause diesel to
a. backfire
b. accelerate
c. decelerate
d. vibrate

218. The volatility of all fuels is described by its_____number.
a. viscosity
b. octane
c cetone
d: saybolt

219. A diesel with a carburetor.
a. opposed piston engine
b. V-8 nonsupercharged engine
c. inline engine
d. none of the above

220. Heat of compression ignites fuel on a gas diesel.
a. true
b. false

221. Most engine heat is released through the
a. cooling jacket water
b. lubricating oil
c. exhaust
d. engine's frame structure

222. Air starting compressors broke down on #1 diesel. #2 diesel is already on line driving a cryogenic oxygen axial compressor. How can you start up #1?
a. tap pressure from compressor to cylinders or air receiver
b. 0_2 pressure is too high
c. 0_2 pressure is too low
d. 0_2 is too cold to flow to diesel
e. none of the above will start the diesel engine

223. What would happen if step 'a' in question #222 was performed?
a. diesel will start if 0_2 can be warmed up
b. diesel can't start 0_2 pressure lifts cylinder safety valves
c. diesel would burn or explode
d. 0_2 will prevent ignition as it is too pure

224. What chemical is frequently used to prevent corrosion in cooling water jackets on a large stationary diesel?

a. sulphate
b. phosphate
c. chromate
d. sulfite
e. caustic soda

225. Compression pressure is normal if found to be 600 p.s.i. on the average diesel.

a. true
b. false

226. Cylinder liners may be classified as

a. dry
b. wet
c. integral
d. all of the above

227. Ambient temperature.

a. exhaust temperature
b. surrounding room temperature
c. blower temperature
d. lube oil boiling point

228. The major cause of diesel shut downs, and poor unstable running.

a. oil lube p.s.i. loss
b. blower failure
c. inferior fuel
d. governor malfunction

229. A diesels compression ratio can be as low as 10 to 1 or as high as 20 to 1.

a. true
b. false

230. Air is taken into a cylinder and exhaust is removed at the same time. What type of engine is this?
a. 4-stroke cycle engine
b. 2-stroke cycle engine
c. 'a' with a blower
d. none of the above

231. Misfiring after assembling engine from a major teardown of inspection and bearing renewal.

a. tight bearings
b. water in oil
c. air locked fuel line
d. cold lube oil
e. 'a' and 'd'

232. A diesel with two camshafts could very likely be an indication of a

a. variable speed engine
b. reversing engine
c. low speed engine
d. combination 2-stroke, 4-stroke

233. When taking over a shift, first check

a. log book and exhaust temperature
b. auxiliary equipment
c. bearing temperature and pressure
d. fuel oil level and engine speed

234. The highest pressure a diesel will have (other than combustion in the cylinder) is somewhat around 3,500 p.s.i. Where will you find this high pressure?

a. cooling system
b. lube oil system
c. fuel system
d. scavenging system

235. When using a indicator diagram, the engine should be

a. at full load
b. cold
c. idling
d. 1/2 normal speed

236. 13 p.s.i. natural gas pressure is okay for a duel fuel engine.

a. true
b. false

237. Start a combination fuel engine on oil, then switch to gas. Switch back to oil when coming off of gas operation when shutting down engine. This will be done with no load on engine and assuming you're running gas fuel at full loads. This procedure also keeps oil pumps primed.

a. true
b. false

238. Vibration can occur if combustion cylinder pressures are not equalized.

a. true
b. false

239. Open cylinder drains before starting because water entry from cooling system could have entered cylinder and will damage engine.

a. true
b. false

240. Engine is properly timed. Knocking is a sure indicator of

a. low fuel pressure
b. overloading
c. air in lube oil near camshaft
d. all of the above

241. A stationary five year old engine constantly overheats.

a. cooling tower fans overspeeding
b. scale in cooling jackets
c. lube oil heat exchanger fouling
d. operating at 50% load
e. 'b' and 'c'

242. Water can be compressed.

a. true
b. false

243. Preignition and overloading is sure to damage big and small-end bearings.

a. true
b. false

244. Oil holes drilled into the crankshaft is no longer used in modern diesels. The 'splash' system is now widely used to lubricate big-end bearings.

a. true
b. false

245. Corrosion on lubricated parts is caused by acid and this acid comes mostly from

a. fuel blowby
b. combustion blowby
c. H_20 in oil
d. all of the above

246. A strain gauge is a dial indicator used to check crankshaft webs for

a. scratched bearings
b. misalignment
c. elastic limit of crank pin

247. Although crank pin is circular it is important that bearing shells are installed with proper alignment to match marks.

a. true
b. false

248. A worn bearing will lower oil pressure due to oil escaping from desired path.

a. true
b. false

249. An important routine test for lube oil.

a. PH
b. H_20
c. particulate solids

d. all of the above
e. none of the above. Change the oil

250. Adjust big-end bearing clearance by using
a. the adjustment screw
b. a new bearing
c. shims
d. a torque wrench

251. When shutting down an engine it is important to shut down cooling jacket water pump immediately to avoid thermal shock.
a. true
b. false

252. Engine efficiency is best at full rated load capacity. 1/3 rated load is inefficient.
a. true
b. false

253. Maintain oil temperature on a large diesel at
a. 50°F
b. 80°F
c. 150°F
d. 200°F

254. Knocking noise from diesels can be a indication of worn bearings and is dangerous to run engine.
a. true
b. false

255. Another term for the needle valve adjustment on a oil relay governor.
a. compensation adjustment
b. speed adjustment
c. speed droop
d. all of the above

256. Protection control on a governor to prevent overloading the engine.
a. brake mean effective pressure limiter
b. governor stop
c. fuel limiter
d. any of the above terms

257. The tie rods of the engine frame
a. transmit main bearing loads into the frame
b. anchor the engine to foundation
c. operate the governor oil pump
d. guide oil to the cam shaft

258. Oil fired diesel uses excess air for combustion to
a. cool cylinders
b. prevent preignition
c. insure complete combustion
d. reduce exhaust gas temperature

259. What causes backfiring in a diesel engine?
a. late ignition
b. leaky valves
c. very lean fuel mixture
d. worn timing gears or chains
a. any of the above
f. 'a' or 'c' only can cause this

260. Threaded hole in the upper piston crown is to
a. prevent carbon buildup
b. cool piston crown
c. install a pressure gauge
d. insert a pull-rod to remove piston

261. Valves bent. What caused this to happen?
a. valve overheated from oil p.s.i. drop
b. timing chain broke
c. valve spring broke
d. piston contact with valve
e. overspeeding engine
f. any of the above

262. How would you know when to change oil or fuel filters?
a. engine will consume more fuel than normal
b. differential gauges will increase
c. when lube oil temperature rises
d. engine will not run at rated speed

ANSWERS - BOILER 500 HP

1.d
2.c
3.c
4.c
5.b
6.b
7.c
8.a
9.b
10*c
11.a
12.c
13*b
14.a
15.d
16*a
17.d
18*b
19*c
20.c
21*b
22. a or c
23.b
24*c
25.d
26*b
27*c
28.c
29.b
30.b
31.b
32*b
33.c
34.b
35.d
36*c
37.b
38.b
39.c
40*b
41.d
42.d
43.d
44*d
45.b
46*c
47.d
48*d
49*c
50*a
51*c
52.c
53.a
54.a
55*b
56*a
57*b
58*a
59.b
60.c
61.b
62.d
63.a
64.c
65*c
66.d
67*a
68.a
69*a
70*b
71*b
72*d
73.b

74.b
75.d
76*d
77.c
78*b
79.b
80.c
81*d
82.b
83.a
84.c
85.c
86*b
87*b
88*c
89.c
90.c
91*c
92.b
93*b
94.e
95.a
96*c
97*c
98*c
99.b
100*b
101.c
102.d
103*b
104.d
105*c
*106.b
107.c
108.b
109.b
110.c
111*c
112*c
113.a
114*d
115*d
116.b
117.b
118.b
119*d
120*c
121*b
122.a
123.b
124.b
125.c
126.d
127.a
128.c
129.b
130*a
131.b
132.d
133.c
134.a
135.b
136*b
137*d
138.b
139*d
140.d
141*b
142*b
143.b
144*c
145.b
146*c
147*b
148*a
149*a
150.d
151.a
152.c
153.b
154*d
155*c
156.a
157.a
158.c
159.c
160.b
161.c
162*b
163.a
164.a
165.d
166.c
167.b
168.a
169*d
170.d
171*b
172*d
173*b
174.a
175.a
176*b
177.d
178*d
179.d
180*a
181.b
182.a
183*b
184.c
185.d
186.e
187.a
188.b
189.a
190.a
191.a

ORAL EXAM TIPS

Never try to fool the examiner by faking a answer. Ask him to rephrase the question if you are uncertain. Or just plain tell him the truth, "I don't know, perhaps we can comeback to this question later as I may recall it."
Never argue with the examiner (incredibly people do this and fail the exam). Tell the examiner how hard you have studied and where you have learned power plant experience. Try to relax and enjoy the ordeal, be cordial and most of all be honest. They know when someone is trying to pull wool over their eyes so don't get sheared by hoodwinking. Be prepared, study, and you will pass the exam. They don't expect you to pass 100%.

Don't drop a resume on the examiner's desk, but a letter from your empl[illegible] indicating dedicated study and satisfactory work performance <u>inside the power plant</u> can be beneficial. Of course, you had better demonstrate your abilities to the examiner otherwise the letter will do you no justice. The letter can also be from an Engineer as a solid recommendation. Keep the letter down to less than one page. Just a couple short paragraphs will do.

Letters are helpful, but how you perform on the exam will determine if you pass or fail. Don't be discouraged if you fail. Many people don't ace the exams until they have taken them a few times. Visit a power plant and ask the engineers many questions...this 'inside information' will be very helpful as they know how to pass exams, and what you need to learn. This is invaluable advice!

LEARNING AND GETTING HIRED

Read all you can.. Attend a trade school holding classes in steam power plant operations or license preparations studies.

Visit as many power plants available in your area Talk to the Engineer's. Ask the Chief Engineer if you can do an 'internship' at his plant. If this is not possible then ask if you can volunteer your time as a employee, full or part-time without pay in training as an apprentice. Using this technique pays off.
1. You are getting to know people in the trade. If the Engineers like you they will want you to be hired when the next job opening arrives.

2. Hiring you is convenient as you already know the plant and minimum time for training will be required.

3. In many cases, a job will be created for you. Most employers will not turn a blind eye away from anyone who shows such an interest in their firm.

4. You don't have to compete with hundreds of other Engineers by going through the long and often frustrating competitive stage of responding to a news advertisement for employment. Who you know is often better than what you know...at least in the 'real world' for getting hired.

The above are excellent methods to obtain employment for the inexperienced. It helps bypass the old dilemma, "Employers want experienced employees, but how do I get experience if they won't hire me?" Now you know how.

For those who already have licensing or job experience you can employ a method I've always used with great success. Never read a newspaper when job seeking, just walk in cold direct to the Chief Engineer's office and talk to him regardless if a job opening exists or not. Don't waste your time talking to the Personnel Manager or filling out an application. Just make an introduction and inquire about and job openings now or anticipated in the future. Ask if he knows of any job openings at other facilities. This is 'networking' and it works!

Often the Chief will know of a job for you or create one for you at his facility. I once told a friend who refused to take this advice because he felt, "That's not how people get jobs." He assumed it is a waste of time to talk to employers who are not hiring in the newspapers. After four months of defeat he decided to give my technique a try. He was hired after only visiting three employers, and he was hired that very same day. The exact results happened to me, many times, even in times of harsh recessions when jobs are scarce. Just walk in and talk about jobs. You'll be surprised at the results you will receive.

Now when you are talking to the Chief Engineer don't talk up a storm about your qualifications and experience. Just act as if you are visiting an old friend. Tell him what you know, then tell him what you want. Then talk about the plant, local news events, etc. Keep the conversation light. This is not a job interview!

The next step is very important. If you approached six employers and all told you there are no job openings don't despair. Do what other people never do, come back in a few weeks and visit your friend once again. Again, keep the conversation light, but do ask if any jobs are available. Most people never return once they discover no jobs are available. If you keep showing up at his door once a month eventually the Chief will help you get a job, or create a position for you.

The above may seem too simple, and unorthodox but it certainly works. And results is what really counts. Right?

Bear these facts in mind. Here is the percentage of people obtaining jobs using four basic criteria.
#1. Employment agencies 11.2%
#2. Want ads 16.6%
#3. Friends and relatives 30.7%
#4. Employer contact 29.7%

That is 84.3 % of people locate jobs using the above four means. The remaining 15.3% remain unemployed. Note that employer contact ranks very high to knowing a friend or relative. Now walking in cold, talking to the Chief Engineer and plant employees builds friendships. You are now using step #3. and #4 simultaneously which gives you a whopping 60.4% advantage than someone using newspaper want ads. Now you know why it works!

LEARNING THE LAWS

As a boiler operator you are responsible for the safe operation of the steam boiler. Remember, it is a bomb! In California it is a Felony for any boiler operator to negligently (a fancy word for accidentally) causing and undue amount of steam to burst from a boiler or engine. Note that the Labor Code law did not say anyone had to get hurt, just making a negligent act to burst steam can get you into trouble. If someone dies from your actions the law has increased penalties. Learn the laws in your state. Obtain a 'umbrella' insurance policy to defend you...just in case an accident happens. Especially if you are a Civil Service employee such as in a correctional facility...you have no immunity as you may have in private industry. Inmates and staff can sue you. Better to be safe than regretful and penniless. Accidents do happen.

LOCATING JOBS

Word of mouth, now called 'networking' is often the best means of locating employment, but what if you don't know anyone in the trade? Large hospitals is a good place to start your search and often the most eager to hire you due to the high turnover rate of boiler operators advancing to higher paid jobs. Talk with the Chief Engineer and visit the boiler plant asking the plant operators if they know who is hiring.

Employment Development Office has a listing, a gigantic book in California, describing all the Civil Service jobs and exams. If you don't see the book, ask for it! You can also visit a state job site and see the listings they have. Visit state hospitals, correctional facilities, jails, government office buildings. They all have bulletin boards and personnel departments to help you find engineering / maintenance jobs.

Contact the local Union. Sometimes they can help you find employment, but not always. Usually you have to find the job first then the Union will sign you on, at least that's how I found it to be in California. At least they can steer you to where the power plants are.

Usually, anytime you see a vapor cloud from condensing steam issuing from a building or you see a smokestack odds are good a steam boiler is present. Even small buildings in cities and states where licensing is mandatory will employ a boiler operator. Check it out!

Of course, electric utilities hire engineers and have apprentice programs. These jobs are 'top of the line' and can be hard to get. Get to know some of the engineers and sooner or later you'll get your foot inside.

If you like weekends and holidays off keep in mind the Civil Service jobs. In California the Dept. of General Services, Dept. of Corrections, Youth Authority all have Stationary Engineer jobs that primarily don't work with boilers. They are maintenance positions and offer very good wages and benefits. Of course, boiler plant positions will not always guarantee day shift assignments.

It doesn't take a great deal of effort to locate power plant jobs. Simply driving around town with an eye open for a stack and vapor cloud will lead you to your first stop. From this point others will direct you where the jobs are. Don't sit around waiting for ads to appear in the newspaper's help wanted sections...the best jobs are often never advertised. Visiting power plants can put you in the right place at the right time.

JOB SEARCH DIARY

Use these columns to log in your job seeking contacts:

FACILITY

____________________.

CHIEF ENGINEER______________.

PHONE___________________.

DATES OF CONTACT:

____
____
____
____
____
____
____
____________________.

CONTACT AGAIN ON:

____
____
____
____
____
____
____________________.

FACILITY

____________________.

CHIEF ENGINEER______________.

PHONE______________________.

DATES OF CONTACT:

____
____
____
____
____
____
____
____
____.

CONTACT AGAIN ON:

____
____
____
____
____
____
______________,

FACILITY ________________
_.

CHIEF ENGINEER________
____ .

PHONE__________
_____ .

DATES OF CONTACT:

____
____
____
____
____
____
____
____
____.

CONTACT AGAIN ON:

____
____
____
____
____
____
______________,

FACILITY ________________
_.

CHIEF ENGINEER________
____ .

PHONE__________
__________ .

DATES OF CONTACT:

____
____
____
____
____
____
____
____
____.

CONTACT AGAIN ON:

____
____
____
____
____

EXPLANATIONS BOILERS 500 HP

1. Sulfite absorbs 0_2. from the air. All tests should be performed quickly.

3. Unless chain operated, the water could burn you. above 200 p.s.i. most water will flash to steam.
8. Miniature bag, bulging, or blister. Only Bags occur in shells.
10. Foaming also produces carryover.
13. Oil under pressure from pump will spray from a small orifice (hole).

16. Stops surging and boiling so we can see the water level.

18. At high circulation rates impurities will not settle.

19. Caused by overfiring which will burn and burst the tube.

21. Farenheit degrees.

24. You are the low water cutout. That is your responsibility.

26. Keep nearby doors closed on cold days.

27. Concentrated acids are often shipped in glass containers.
32. To prevent priming when valve lifts obstructing steam exit path.

36. Clean out; boil unit under low pressure with chemicals.

40. Anything that will not 'draw' steam.

44. With main stop valves closed completely.

46. Be careful as chipping will take brick work with it.

48. With steam or compressed air.

49. For blowing tubes.

50. Punching a tube is referring to wire brushing firetubes.

51. A valve when opened you can see straight through it.

55. Placing one or more boilers on an existing pressurized main steam line.

56. To prevent disturbing steam formation, and prevent thermal shock.

57. Open drain to glass to relieve pressure. Make sure you do this!

58. This practice is dangerous. Don't do it! Establish a pilot or torch

65. Or not enough air for fuel to burn.

67. Flue is the ductwork connecting boiler to the stack.

69. Boiler tubes can be corroded also at low firing rates.

70. Someone may get killed if workers are inside boiler. Even with valves closed don't blow down any other boiler until safe to do so.

72 Many are designed not to close completely.

76. Chlorides are salts.

78. Air will build up a false pressure temperature relationship. This will cause thermal shock. Forcing the fires to compensate when putting a boiler on the line, air leaves quickly with no steam to fill void. Proper temperature will not be reached.

81. Some are beaded for drainage on lower steam and water drum and upper mud drum to remove entrapped air accumulation pockets.

86. Two gage glasses tapped into drum is Okay. See the A.S.M.E. code for details.

87. For drainage, blow down valves are in the rear.

88. Attached to upper shell to upper head.

91. Valve 'pops' open.

93. 1/16 inch lower because water is cooler and denser outside of the steam drum.

96. So impurities will not carry from column to obstruct glass connection.

97. Use ultraviolet if detector sees glowing furnace slag with flame out.

98. Feed pump takes care of pressure and recirculation systems.
100. Steaming is not to be allowed. There is no thermal cooling circulation in pipe.
103. This is what activates regulator via sensor lines.
105. Steam volume is larger. Higher pressure needs only a smaller valve.

106. Steam condensing. Water level rises in drum, unit is no longer generating steam. Feedwater valves should be closed and not leaking.

111. Steam is slightly superheated due to exposed firetubes to steam.

112. Stay bolts.

114. If plates are hot, water will flash violently if introduced. Steam generated capacity will exceed safety valves capacity = explosion.

115. To thoroughly mix fuel and air.

119. Open glass drain, close steam connection then open. Close water connection then open. Close glass drain. Try this out for yourself. By closing the steam connection you are blowing the water connection and vice-versa.

120. In case of a boiler tube failure, the whole system will back up and dump steam into this unit.

121. Demineralized water.

130. Helps distribute water evenly to prevent thermal shock.

136. Deposits in tube will cause overheating and failure.

137. Unless tandem hard seat valves or seatless valves are used, then it will be the blowing valve on the inside.

139. When operating, crack open valve. Steam has a blue tint color and emits a high pitch sound. Water flashing to steam and condensing has a white cloud appearance with a softer 'wispy' sound.

141. To prevent blowing the flame out.

142. Nucleate boiling = steam film layer in tube keeping water from tube.

144. Answer 'd' wrong. Don't over pressurize steam headers and auxiliaries.

146. Pertaining to lower ring. If you raise upper ring blow back will decrease. Study your safety valves. They are the most important to know.

147. Some are round, usually below 100 p.s.i. and above 350 p.s.i.. Depends on metals used to manufacture spring and temperature service rating.

148. If pump is running, let it run unless steam pressure rises abnormally.

149. Answer 'b' is OK, but it's not wise to close a feed valve. Blowing down is economically wasteful.

154. Boiler is not designed to work in reverse stress, it <u>could</u> collapse.

155. It can be possible in third pass if heat rate is high, but unlikely.

162. Remove the condensate out so you won't impinge tubes.

169. Soot on heating surface of tube will not cause failure only fuel efficiency loss.

171. Answer 'd' is correct, but will not 'cure' foaming.

172. Shutting down fires is usually sufficient depending on the unit operating characteristics.

173. Supports brick walls. Usually an 'I' beam.

174. Lap joints are found on old rivet steam & water drum shell sections.

175. A stronger riveted joint.

176. Slagging is carbonized oil on furnace structure.

177. Shells connect tubes and have many holes so they must be thicker than tubes.

178. Answer 'c' is most prone.

179. No explanation needed.

180. Creates a density difference.

181. Answer "b" is correct if you still see water in the gage glass.

182. Acids can do the same. Carbonic acid from carbon dioxide.

183. Excess air in any amount is wasteful, but some is necessary to accomplish complete combustion.

184. As pressure builds it forces steam bubble to compress lowering water level in drum.

185. These are the primary gases. Others do exist.

186. Electronic sensor monitoring is now used for best results, if calibrated.

187. Each fuel has unique flue gas ratios.

188. Warming and cooling a boiler slowly are both critical to avoid stress.

189. Glass will break if touches metal.

190. It should 'ping.' If it 'klunks' you know it's broke.

191. Draft gages will tell you by how much to adjust dampers.

ANSWERS - AUXILIARIES 500 HP

1.b
2.d
3.b
4.c
5.a
6.c
7.d
8.b
9.a
10*d
11.d
12.d
13.b
14*a
15*d
16*b
17.a
18.b
19*a
20.c
21*b
22.c
23.c
24.b
25.a
26.b
27*c
28.b
29*b
30*b
31*d
32*d
33*c
34.d
35*b
36.c
37.d
38.d
39.a
40*d
41*a
42.a
43.a
44*b
45*a
46.a
43*a
44.b
49.a
50.b
51*b
52*a
53.a
54.b
55.c
56*b

EXPLANATIONS AUXILIARIES 500 HP

1. Actually increases steam flow.

2. Cannot see steam or air.

3. A safety relief valve performs 'a' then 'b.' Safety's always "pop" open wide and stay open until blowdown is accomplished, then closes quickly.

6. Cyclone is the key word to arrive at answer 'c' but can be found on 'a' 'b' and 'd'.

9. Controls returns to receiver and steam to feed reciprocating pump.

10. A vent condenser is not called an ejector. Air eliminators can be installed on DA's and can be called an ejector. 'd' is a complete answer and the best choice.

14. And may come in

'b' 'c' or 'd' form.

15. 'c' on trap

discharge.

16. Elements refer to as what the regulator senses to operate. In this specific case water level in steam and water drum = 1 element.

19. Coils or tubes: Shell & tube heat exchanger.

21. As water expands gasses are liberated. Answer 'c' can also accomplish deaeration.

26. A blow-off tank is a flash tank. If flashing exist in receiver, traps are blowing steam.

27. Thermostatic float trap is used when steam is condensing due to air in steam.

29. To prevent steam entering sewer and scald any workmen who may be performing repairs.

30. Flue = sheet metal connecting boilers to stack.

31. Reduces thermal shock to 'c'.

32. Fluid acting on all faces of valve disk = balanced.

33. Remove impurities from zeolite beds.

35. See introduction on terminology.

40. Valve seats rust or become locked together.

41. This will reduce velocity and eliminate noise.

44. Flowing steam at low pressure has cooling effects.

45. To force returns against atmospheric resistance in return line. Returns is a term for condensate returning to receiver.

47. Will not develop full power, but enough to cause reverse power conditions when driving A.C. generator. This will cause damage to generator and turbine if you close the breaker in this situation. Turbine oil pump will not pump oil if shaft driven = damaged bearings.

51. Blow-back adjustment = too little blowdown. Blowback and blowdown means the same, just different terminology.

52. If located to blow from front head installation flame must be shut down.

56. Viscosity too low, gears wear from contact.

ANSWERS COMPRESSORS 500 HP

1. b
2. a
3. b
4. b
5. c
6. b
7. b
8. d
9. c
10.b
11.b
12.e
13.a
14.a
15.a
16.e
17.d
18.a
19.c
20.a
21.a
22.e
23.b
24.d
25.e
26.b
27.b
28. Shut down.
29.b
30.e
31.b
32.a
33.b
34. Pressure can't return to low side.
35. Inlet valve leak.
36.a

ANSWERS - MECHANICS 500 HP

1.c
2.a
3.b
4.d
5.d
6.a
7.d
8.c
9.c
10.a
11.e
12.b
13.a
14*a
15* none
16*d
17*c
18.e
19.a
20.a
21*a
22*c
23.a
24.b
25*a
26.a
27.d
28*b
29.a
30.b
31*d
32*b
33.b
34.b
35.a
36*b
37*b
38*b
39.c
40.a
41.a
42.b
43.a
44.b
45*b
46*a
47*b
48.a
49*c
50.a
51.a
52.d
53*d
54*a
55*a

EXPLANATIONS - MECHANICS 500 HP

2. A straight cut gear is a 'spur' gear.

3. Any pump that must lift fluid. Lift always pertains to suction. Draft head pertains to discharge or suction under positive pressure.

6. Nitrogen is inert = non reactive, 02 could cause a fire.

7. This is a trick question. Answer 'd' implies that Teflon can be used on all of the above.

14. A pigtail is needed only to prevent heat damage to gauge from acting fluid.

15. 90w is liquid, just pour it into casing.

16. If noise persist, shut down.

17. This is a winding fault and will catch fire.

21. If oil coil or tube ruptures, the oil will back up into boiler.

22. A forged steel needle type globe is preferred.

25. Non-return is a globe with 'straight-away' properties.

28. You may adjust valve, but results must be inspected by the insurance company or

inspector. It is best policy to do the adjustment in his presence.

31. A special tool is used.

32. You can be held responsible if accident occurs.

36. Scratches weaken tensile strength in any location. Glass will fail.

37. Double extra heavy.

38. W.O.G. = water oil or gas (not for use with heated fluids at high temperature).

45. Never get electrical equipment wet. Clean windings by blowing with low p.s.i. nitrogen.

46. To avoid picking up sediment.

47. Answer in relation to clock dial.

49. 'S' stands for saturated steam service.

53. Dry packing will wear shaft.

54. Soap stone to be soaked in oil.

55. If you're holding a license to operate, you could lose the license and be subject to fines or imprisonment. Know the laws in your state.

....

ANSWERS - PHYSICS 500 HP

....

2.c
3.d
4.b
5.a
6.a
7.d
8.a
9.c
10.b
11*c
12.e
13.c
14*c
15.c
16*a
17.b
18.d
19*b
20.c
21.c
22*c
23*b
24.c
25.d
26*d
27*c
28*b
29.a
30*a
31*a
32*a
33*a
34.b
35.d
36.b
37.d
38.d
39.d
40.a
41*a
42.a
43.a
44*b
45.c
46* 'a' 'b' 'c' 'd'
47.b
48.a
49.a
50.d
51*a
52*c
53*b
54*c
55*e
56.d
57.c
58.c
59.a
60.d
61.d
62*b
63.a
64.c
65.c
66.c
67.b
68.c
69.d
70.a
71.c
72.c
73.a
74.b
75.c
76*a
77.a
78.a
79.d
80.a
81.b
82.a
83.a
84.c
85*e
86.b
87*a
88.c
89.a
90.a
91.a

EXPLANATIONS - PHYSICS 500 HP

1. Exhaust can also be high pressure (see Physics Unlimited Question No. 50 and 51).

6. Vacuum or heat. Pressure is irrelevant without heat.

7. Remember this! Can you see steam in the gauge glass?

11. Nitrogen will produce no heat. It is an inert moderator to slow down combustion.

14. Go look at the barometer scale to cure any doubt.

16. Removing heat also applies.

19. Actually 'a'-'b'-'c' is also correct for making calculations.

22. Radiation travels in straight lines. It is independent and will not convect.

23. Answer 'a' is only correct at sea level.

26. Atom loses or gains an electron is ionized.

27. Under atmospheric conditions.

28. Atmospheric pressure.

30. Converge = compressing, come together.

31. Diverge = expanding, spread apart.

32. Think! Deareator, a receiver, will be a false answer. (see #34).

33. Steam entering water will also implode causing hammer effects.

41. More elements are being discovered in nuclear research.

44. Oil in steam reduces quality. Wet steam is common from engines due to cut-off and expansion in cylinder.

46. On the acid side of P.H. scale = 'a'-'b'-'c'-'d' are all correct as they all can be acid.

51. Pressure in relation to head.

52. Via hydrogen ion content.

53. Nothing removed, only changed.

54. Nitrogen retards combustion, although an inert gas (non-reactive), at temperature will form, for example; oxides of nitrogen.

55. Hydrazine also absorbs 0_2.

62. Actually, to be technical 'a'-'b'-'c' is correct.

76. Pure water is $H_2 0$. Most impurities in water will require a formula that is not feasible here due to the complex variables.

85. Also applies to #84. Fluid exerts uplifting force.

87. Alternating current.

ANSWERS - PUMPS 500 HP

Don't cheat! You only hurt yourself. Make sure you answered the question before looking up the answer. Try as hard as you can to pass the test without resorting to looking up the answer. Resist the temptation to take shortcuts...if you plan on memorizing the answers in this book it may do you little good in the long run.

Make sure when you do look up the answer that you learn <u>why</u> you marked it right or wrong. Simply looking up the answer alone will not benefit you as you will sooner or later forget or become confused.

Discus the answers with a engineer at a power plant. He can explain in detail why the answer is right or wrong and show you the item in the power plant in question. After all, seeing is believing because learning power plant engineering isn't a feat of magic, but of logic based on the laws of physics.

1.b
2.b
3.c
4.d
5.d
*6*b*
7.c
8.c
9.a
10.b
*11*b*
*12*d*
13.e
14.a
*15*b
16.d
*17*d
18.d
*19*b
20.a
21.b
*22*c
*23*b*
*24*d*
25.a
*26*b*
27.e
28.c
29.d
30.b
31*b
32.b
33.c
34*d
35.d
36.b
37.c
38.c
39.a
40*b
41.b
42.d
43.e
4 4.a
45.b
46.d
47.d
48.d
49.b
50.a
51.b
52.c
53.b
54.d
55.a
56.c
57.a
58.d
59.e
60*b
61.b
62.b
63.b
64.d
65*b
66.a

EXPLANATIONS - PUMPS 500 HP

1. This prevents water hammer shocks.

6. Fluid whip creates heat. Can be used to maintain a set head pressure.

7. Close vent after starting, open discharge slowly at line pressure. This is very important to know.

9. Very small lap. Only enough to cover ports.

11. No opposing force on active medium (piston, impeller). Check valve in discharge line takes the load off pump (if it doesn't leak). Unloading also prevents overloading driver. C.V. stops reverse flow to deareator.

12. Steam imparts energy and heat to fluid. Ideal for boiler feed.

15. Close vent when you see steady stream of water.

17. 'b' is correct if area is changed to diameter.

19. Head = energy in fluid. Related to pressure. Motion develops head not pressure.

22. Suction line is always larger than discharge.

23. Opens according to pressure rise.

24. If lifting fluid, a check (foot valve) will be found and a strainer.

26. Bearing and / or wearing ring damage.

31. See: 'Pumps Unlimited' section to find out why.

34. 'a' is not correct due to pump suction under positive head for boiler.

40. Air bound pump will build no pressure or very little. The noise in a steam bound pump is steam imploding, expanding or compressing.

60. This is a special fitting to prevent air pockets in suction line and to help smooth the inlet flow of fluid to pump = reduce turbulence.

65. CO_2 is correct, but dry ice relates to this question.

ANSWERS - ELECTRICITY 500 HP

1.a
2.c
3.d
4 *a
5*a
6*c
7.b
8.b
9.a
10.a
11.a
12.a
13.b
14.d
15.c
16.b

EXPLANATIONS - ELECTRICITY 500 HP

4. Large bussway feeders are multi-stranded for flexibility.

5. Resistance decreases with temperature decrease.

6. Any metal conducts. Air also, as in lightning arcs or slip ring flashovers or phase to phase arc jumps.

ANSWERS - TURBINES 500 HP

1. c
2.a
3.d
4.c
5.d
6.b
7.a
8.c
9.a
10.b
11a
12.b
*13*e*
*14*e*
15.d
16.c
17.b

18.b
*19*a*
*2 0*c*
21.c
*22*a*
23.d
24.c
25.d
26.b
27.a
28.c
29.a
30.b
31.a
32.b
33.e
34. e
35. d
36. c
37. b
38. b

EXPLANATIONS - TURBINES 500 HP

2. Properly called a hydraulic turbine.

3. If for a strange reason cooling water is condensate it could get into boilers. If circulating water is used, tower will be contaminated if 'b' were to happen.

4. Any gas will run a turbine even hydrogen, but is dangerous if 0_2 entered turbine. Many inflammable gasses are expanded in turbines to act as a reducing valve extracting "free energy" in the process i.e. natural gas.

6. Steam strikes at various design angles and is deflected, but we will use answer 'b' so you will understand the basic theory.

9. Speed will not lower. Governor will compensate.

13. All steam contains moisture even live steam, although exhaust steam has more, depending on heat content.

14. Putting steam to the seals guards against air leaking into turbine. Not for lubrication, although, steam has a cooling effect on the seal.

19. Stress will throw alignment out.

20. Assuming hertz indicator is accurate.

22. Hydraulic turbine is also the same as in question #21 answer 'c'.

INTRODUCTION TO UNLIMITED

ENGINEERS' EXAM WITH TURBINE ENDORSEMENT
based on Los Angeles California requirements.
Requirements subject to change.

If you wish to obtain only the Unlimited Boiler Operators license, leave out the turbine section. In any case, you must complete the 500 horsepower exams before you can qualify to attempt the unlimited test as the contents are much more involved.

Steam engines are in use today and are still manufactured. We will not explore in-depth steam engines. I do appreciate any correspondence and suggestions you may have for future editions.

The following is a brief example of what you can expect when taking an oral examination.

TYPICAL ORAL EXAM

1. Trace water through a boiler pressure system with a condensing turbine and include the ion-exchangers.

2. Give me 3 ways to control superheat. Mention 1 more.

3. Describe the chemicals used at your plant and purpose. How do you stop condensate return line corrosion? What chemical is used?

4. What is the main function of a condenser to turbine relationship? How do you start and stop one?

5. Taking a turbine off the grid, what would you do to prevent the rotor from warping?

6. What is the A.S.M.E. code pertaining to safety valve blow back? How do you determine safety valve capacity?

7. What can cause a turbine to trip? Give me more, as many as you can.

8. Why does a steam pump's (duplex type) pump against boiler pressure that takes its steam pressure from the same boiler?

9. How do you remove 0_2 from feedwater?

10. What is a inspirator? Injector? Ejector? Tell me how to start them up.

11. What is a static and dynamic head referring to fluid pumps?

12. What is lap and lead referring to? Do steam pumps have it? Set the valves? (on a steam engine and pump) On a steam engine what is the angle of advance? On a steam engine, twice the radius of the throw of the eccentric is what?

13. How would you start a vertical fire tube boiler? Is it internally or externally fired?

14. Give me five ways to introduce feedwater into the boiler.

15. If you were burning oil and getting no returns (as in a 1 pass system) and you got oil in the boiler-where did it come from? And how would you prevent it from happening again?

16. Describe latent heat, specific heat, sensible heat.

17. If you were running with turbine driven forced and induced draft at peak load demand and you spun a bearing on the forced draft fan, what would you do? Now tell me what you would do if the induced draft fan bearing failed?

18. If you left the boiler only to come back to find the boiler shut down and no water in the gauge glass, but the feed pump is still running dry, what would you do?

19. Start a duplex steam pump with a leaky check valve on the boiler.

BRIEF ANSWERS TO ORAL EXAM

1. Draw boiler, steam line to prime mover, condenser, condensate pump and line to deareator. Now continue drawing suction line from DA to a boiler feed pump, through a closed heater, economizer, then to the boiler.

The three ion exchangers, you may find two locations for them. A demineralizer (condensate polisher) will be found between condensate pump and the deareator. A demineralizer and water softener will be found on makeup line to deareator.

The examiner could ask you to install expansion loops and various valves and bypasses, you should know this blind folded. Practice drawing as much equipment as you can on your basic steam and water cycle, it will help you pass exams.

2. Steam pressure, firing rate, dampers, flue gas recirculation, attemperation.

3. Phosphate, sulfite, in boiler. To combat return line corrosion, amines control condensate alkalinity limits. Hydrazine also absorbs oxygen.

4. Condenser produces vacuum to reduce atmospheric resistance to turbine exhaust producing power and increased efficiency. To start - circulate condenser water, start air ejectors (hoggers), start turbine.

5. Place turbine on turning gear. (grid = electrical load term or powerline distribution system used in electric utilities).

6. Not less than 2%, not more than 4%. Close stop valves, run fires to maximum, pressure on steam gauge shall not exceed 6% of maximum allowable working pressure. If pressure rises beyond 6%, shut down, reduce burner capacity or add larger safety valves. Always perform test with water level at two gages or less.

7. High and low oil pressure, vibration, overspeed, vacuum loss in condenser, any solenoid such as cooling water flow interruption, oil temperature, etc. (see Turbine notes) Also, remember if a trip occurs and extraction check valves don't close, turbine will overspeed and may most likely induce water from feedwater heaters (just something to think about).

8. Area of steam piston is 1 to 1 1/2 times the area (larger in size) of water piston = increased total force applied.

9. Deaeration, sulfite chemicals.

10. Manually operated feed pump with forcer valve and two overflow valves. Forcer valve operates the overflow valves internally. No valves. All three pumps work by diverging nozzles and converging nozzles. Lift is created by the diverging nozzle via vacuum effects. Converging nozzle imparts head pressure to the water from flowing steam pressure. Ejectors do not pump against pressure, but the other two will.

11. See Pumps & Physics. Static - is usually referred to suction line pressure with pump stopped, but can be referred to pressure acting on discharge check valves.

Dynamic - is related to both suction or discharge fluid pressure in motion.

12. Slide valves in steam engines. No. To set engine valves, tram engine to find dead center. Rotate 90° plus 10° lap and 10° lead. To set pump valves - center pistons in mid-travel of cylinder, plumb rocker arms, square valves and equalize lost motion. Angle of advance is 110° . Twice eccentric radius = eccentric throw or travel.

13. Full of water. As steam forms, blow down. Do this in stages. Internal.

14. See Pumps.

15. Check valve on steam line to heater leaking (or not even installed) so when oil heater tube ruptured oil went right into the boiler. Usually happens when starting up or shutting down boiler. Install or repair check valve and heater.

16. Latent heat creates a change of state. Sensible is thermometer readings, specific is total heat content of any matter. For instance: 1 lb. of iron and 1 lb. of water both at 200° F. The iron absorbs and retains much more heat energy than the water due to the differences in mass.

17. Speed up the induced. Speed up the forced.

18. Nothing. If you put water on hot plates the steam generated will flash quickly beyond the safety valve capacity. Most certainly exploding the boiler.

19. Close discharge valve, open vent, open drips, crack steam valve, oil pump's moving parts, close drips when steam exits. Bring up speed with steam supply valve. When water pressure equals or exceeds boiler pressure close vent and open discharge valve.

ODDS & ENDS

A tandem compound turbine - two turbines set one behind the other with rotors connected. The high pressure turbine exhaust into the low pressure turbine.

A cross-compound turbine is two individual turbines parallel to each other, but high pressure turbine exhausts into low pressure turbine, of course, rotors are not connected to each other.

All superheaters are not always in the boiler furnace, so called 'external superheaters.' Some are oil or gas fired pressure vessels and others are heat exchangers which superheated steam can be used to superheat steam. Of course, the former must be hotter than the latter, unless desuperheating is to be desired. <u>(also see</u> Auxiliaries #43)

Centrifugal oil relay governor - the centrifugal weights sense speed changes and transmits changes via plunger to a oil pilot valve that admits or bleeds oil to the power piston connected to the throttle linkage. This is the combining of two governors and not to be confused with the application of a centrifugal governed (fly-ball) engine or a engine with a oil relay governor.

Dummy Piston - (balancer) <u>is</u> simply a piston or surface that is used to counter the axial thrust forces of a rotor. Picture in your mind a large condensing turbine. The thrust force is traveling toward the low pressure end loading the thrust bearing to maximum. Imagine placing a disk on the rotor just before the thrust bearing and putting high pressure steam behind this plate (piston) to counter the thrust force in the opposite direction away from the thrust bearing. Steam is pushing one way and now we have steam pushing the other way to balance the thrust force. If this was not done, thrust bearings wouldn't last very long. Remember, large centrifugal pumps have balancers using pumped fluid to counter balance thrust. Can you see the advantage of an opposed-flow turbine?

Changing over pumps. (See 500 Horsepower question #1-Pumps). Supposing you didn't perform the proper procedure, what could happen? If you trip one pump you may not get the other started in time due to a 'leaky' check valve. If the check valve (s) slams shut and breaks (with a centrifugal pump) the water in the boiler with all the heat energy with it will back up to the deareator and flash violently over pressurizing and blow up the deareator.

Safety valve on DA will not handle this volume. DA - M.A.W.P. is likely only 50 psi.! So always think what could happen before doing any task, for it just might happen.

Many old gaskets, insulation and packings were made with asbestos. Today, hazardous specialist are used to remove asbestos. For small repairs which produce little dust such as scraping a gasket flange, you should wear a fine micron dust mask. Asbestos does cause cancer. Don't remove old pipe insulation as you may be held liable for exposing employees to asbestos. Today, many substitutes for asbestos exist, although they do have temperature limitations.

ANSWERS - AUXILIARIES UNLIMITED

1.a
2.d
3*a
4*b
5.b
6*d
7.c
8*d
9.a
10.b
11*a
12.b
13.d
14.f
15*c
16.e
17.b
18.c
19.f
20.a
21*a
22*d
23*d
24.d
25.d
26*c
27*d
28.d
29.d
30.b
31*b
32.d
33.a
34.d
35*c
36.d
37*d
38.e
39.d
40*d
41.e
42.d
43.d
44.d
45*e
46*c
47.a
48.c
49.a
50.b
51*a
52*a
53.a
54.a
55*c
56*a
5 7*a
58*a
59.c
60.a

EXPLANATIONS - AUXILIARIES UNLIMITED

3. All dissolved gasses present.

4. The key words to watch in this question, "in the." 'a' and 'c' is where the test "is" taken or may be taken.

6. Anti-foam chemicals are used if foaming is a problem.

8. From water.

11. Steam space has been reduced. Nozzles or drip pans are flooded.

15. If boiling takes place, you will get false reading and fluctuations.

21. This is a trick question as most will mark answer 'b' as correct, but 'a' actually exists. Check with a trap manufacturer.

22. Designed for high capacities. Thermostatic float trap is best. Do you know why?

23. I f this happens, then you will see 'a'.

26. All traps other than 'c' are classified as separating traps.

27. 'c' closes feed valve. All returns now flood DA. Remember this when you ever trip a boiler out.

31. Never trust alarms as they can fail.

35. Some deareators vents have a valve to control steam bleed (see #39).

37. Sentinel relief valve is a better name for this safety valve. Do you know why?

40. A vacuum will not form bubbles. Fluid will be under suction.

45. Fluid medium = circulating water

46. 'c' is most important. Strainer before valve, check valve after valve if backflow conditions exist, depending on the valves service.

51. Corliss has a trip latch on valve stem.

52. Some examiners may want for an answer; "the volume of water delivered to suction controls speed of a pump."

55. This will result in distilled water putting steam to a closed heater will increase plant efficiency and will not upgrade quality of feedwater, but is good to know for energy conservation purposes.

56. 'b' is wrong as condensate pumps to DA, suction then piped to feed pump.

57. Catchy question! Injectors cannot pump hot water. Even if they could, why install injector is such a situation that DA could rupture or feed line break rendering injector useless?

58. Diesels have injector racks and could come close to shaft governor term, if it ran off of flywheel. Very close resemblance and debatable as governors do run off of flywheel. This one is your choice.

ANSWERS - BOILERS UNLIMITED

1.b
2.a
3.d
4.b
5.c
6*b
7.a
8*b
9.b
10.a
11.b
12.b
13.b
14.c
15.c
16.a
17.c
18*a
19*a
20.b
21.b
22.d
23.d
24.d
25.a
26.c
27.b
28.a
29.c
30.b
31.c
32.d
33.c
34.b
35.b
36.b
37.b
38.b
39.e
40.d
41.d
42.a
43.b
44*a
45.b
46.b
47.b
48.d
49.b
50.a
51.b
52.d
53.a
54.c
55.c

56.b
57.d
58.c
59.b
60.a
61.c
62.c
63.a
64.b
65.d
66.a
67.b
68.a
69.d
70.a
71.b
72.a
73.a
74.d
75.c
76.c
77.b
78.a
79.d
80.d
81.d
82.a
83.d
84.a
85.b
86*a
87.b
88.e
89.d
90.a
91.b
92.c
93*a
94.b
95.d
96.a
97.b
98.d
99.d
100.a
101.a
102.b
103.d
104.d
105.c
106.a
107.a
108.b
109.d
110.a
111.b
112.a
113.b
114*c
115*a
116.c-d
117.b
118.b
119.c
120.b
121.c
122.c
123.c
124.a
125.d
126.b
127.c
128.b
129.c
130.a
131.a
132.a
133.b
134.b
135.b
136.d
137.b
138.a
139.b
140.b
141.b
142.b
143*b-a
144.a
145.a
146.b
147.a
148.a
149.d
150.a

151. See 1 through 10 below.
1 matches j
2 matches i
3 matches a
4 matches f
5 matches e
6 matches b
7 matches g
8 matches h
9 matches d
10 matches c

EXPLANATIONS - BOILERS UNLIMITED

6. At least.

8. Turbine is the obstruction or any isolation valves.

18. If inspections have passed.

19. Sometimes both sides.

44. Too much slag could cause 'b'.

86. Dissimilar metal contact = electrolytic corrosion.

93. Impurities will insulate tubes then chip off to enter turbine.

110. This causes the water to bounce into the glass if present.

114. Boiler load increases, feed pump vacuum increases water flashes. 'c' is also a problem. Do not open to line capacity.

115. Keep boiler water level at 2 gages to prevent this.

141. When pressure is zero and vent no longer emitting steam, drain water, open manholes and spray high pressure water to flush deposits.

143. 'c' is wrong. Fire tubes are always under compression.

ANSWERS - COMPRESSORS UNLIMITED

1.a
2.b
3*a
4.d
5.d
6.a
7.d
8.b
9.d
10.d
11.c
12.a
13.a
14.a
15*a
16.c
17*b
18.b
19.b
20.d
21.c
22.a
23.a
24.a
25.a
26.a
27.d
28.d
29.d
30.e
31.b
32.e
33.e
34.a
35.a
36.a
37.d
38.c
39.b
40.c
41.c

EXPLANATIONS - COMPRESSORS UNLIMITED

3. Dew droplets = improper terminology. Also, condense oil vapor.

15. Vanes are automatically unloaded due to inherent design features.

17. If they do, it is only slight spring tension.

ANSWERS - STEAM ENGINES UNLIMITED

1.d
2.a
3.a
4.c
5.c
6.d
7.b
8.e
9.b
10*b
11.c
12.a
13.a
14.d
15.a
16.d
17.b
18.b
19.a
2 0*b
21.d
22.d
23.d
24.b
25.b
26.a
27.c
28.c
29.d
30*c
31.c
32*c

EXPLANATIONS - STEAM ENGINES UNLIMITED

10. From this point you may then find 'a' and 'b'.

20. Counterflow engines do have variable cut-off, but not with slide valves as most slide valve engines have "fixed" eccentrics

30. Oil-filled piston like a shock absorber.

32. Stranded cable is stronger than solid stock in tension, but flexibility is the reason for stranding.

ANSWERS - PUMPS UNLIMITED

1.b
2.c
3.a
4.a
5.c
6.c
7*d
8.a
9*b
10.b
11.c
12.b
13.a
14.b
15.b
16.c
17.a
18.b
19.d
20.c
21.b
22.b
23.a
24.a
25.a
26.b
27.d
28.c
29.c
30.d
31.a
32.e
33.d
34.a
35.g
36.c
37.a
38.a
39.c
40.a
41.b
42*b
43*a-b-e- f

EXPLANATIONS - PUMPS UNLIMITED

7. Screw pumps are of the rotary class, but are called

rotary screw.

9. Due to the water side passages.

42. Small pumps have a rotor locating bearing not normally
called a thrust.

43. Return traps for low p.s.i. boilers only. Very few in use today.

ANSWERS - TURBINES UNLIMITED

1.b
2.a
3.a
4.c
5.a
6.b
7.b
8.b
9.c
10.b
11.b
12.d
*13**a
14.c
*15.*b
16.b
17.b
*18.*d
19.c
20*c
21.d
22.c
23.a
24.c
25.c
26.b
27.b
28.a
29*c
30.d
31.b
32.d
33.b
34.c
35.d
36. a
37.d
*38.*a
39.a
40*c
41.a
42.b
43.c
44.b
45.a
46.d
47*a
48*b
49.c
50.b
51.c
52.b
53.d
54.b
55.a
56.d
57.b
58.d
59.c
60.a
61.d
62.d
63.a
64.a
65.b
66*d
67.a
68.b
69.c
70.d
71.b
72.d
73.d
74.a
75*b
76.e
77.b
78.b
79.a
80.a
81.b
82*b
83.b
84.c
85.d
86.d
87.d
88.b
89.a
90*b
91.b
92.c

93.a
94.a
95.d
96.a
97.e
98.b
99.a
100*a
101.a or c
102.e
103.b
104.c
105.a
106.e
107.e
108.a
109.d
110.d
111*a
112.b
113.d
114.b
115.b
116.a
117.b
118.c
119*b
120.a
121.d
122*b
123.c
124.a
125.a
126.a
127.d
128.d
129.a
130.a
131.d
132.c
133. a or b
134.b
135. See 1 through 10 below.
1. matches b.
2. matches d.
3. matches c.
4. matches a.
5. matches h.
6. matches g.
7. matches e.
8. matches f.
9. Matches j.
10. Matches I.

136. d

EXPLANATIONS - TURBINES UNLIMITED

3. Steam strikes blades on a greater angle than reaction.
5. 'b' is correct if excess steam is condensed to regulate pressure in exhaust system.

6. All steam condensed to produce vacuum.

13. Axial clearance.

20. Does not require oil to function, only for lubrication.

29. Double seated valves produce balanced effects.

40. Sealant and string sealant is used.

47. 'b' should already be in process.

48. #53 also applies. Thermal shock = unequal expansion.

66. Drains are piped to condenser on a condensing turbine.

75. 'a' can cause this too.

82. Rotor warping due to its weight.

90. Even the air we breath is superheated.

100. Term used to turn flywheel on steam engines.

111. Heat absorbed and utilized = heat rate cycle efficiency.

119. Oil relay via linkage. Shaft governors are on steam engine flywheel.

122. Exhaust flow and pressure reduced allowing water in turbine.

ANSWERS - PHYSICS UNLIMITED

1.a
2.b
3.c
4.b
5.c
6.a

7.c
8*a
9.a
10.e
11.a
12.b
13*b
14.a
15.a
16.a
17*d
18.d
19.a
20.c
21.a
22.c
23.a
24.a
25.a
26.b
27.a
28.a
29.a
30.d
31*b
32.b
33.a
34*b
35*b
36.c
37.c
38.a
39.a
40.a
41*b
42.a
43.a
44.a
45.a
46.c
47*a
48.b
49*c
50.a
51.a

EXPLANATIONS - PHYSICS UNLIMITED

8. This applies to solids also.

13. Sulfur and water = sulfurous acid.

17. Depending if heat is added or removed.

31. 1 bar = 1 atmosphere pressure.

34. Temperature is not high enough to isolate atomic structure. Steam will form and use up heat = a waste of fuel.

35. Friction.

41. Answer #41 and #42 is not contradictory. Learn the difference.

47. Enthalpy calculation will prove 'a' to be correct.

49. 100 p.s.i.a. = 327° F. - 100 p.s.i.g. = 316° F.

ANSWERS - REFRIGERATION UNLIMITED

1.b
2.a
3.c
4.a
5.c
6.c
7.e
8.d
9.a
10.b
11.d
12.b
13*a
14.a
15.d
16.a
17*a
18.d
19*d
2 0*c
21.d
22.c
23.d
24.b

EXPLANATIONS REFRIGERATION UNLIMITED

13. All gasses will displace oxygen. Although not toxic by itself, will cause asphyxiation.

17. Partial expansion.

19. Used in absorption systems as an example.

20. Strong concentrated salt. Irritating to the touch.

ANSWERS - MECHANICS UNLIMITED

1.b
2.a
3.b
4.b
5.b
6.c
7.a
8.c
9.d
10.a
11.b
12.a
13.d
14.b
15.c
16.a
17.d
18.c
19.d
20*e
21.a
22.a
23.b
24.a
25.d
26.a
27.d
28.b
29.a
30.b
31.a
32*b
33.e
34.b
35.b
36.a
37.a
38.d
39.a
40*b
41.b
42*c
43.a
44.d
45.b
46.b
47.b
48.b
49.c
50*c

EXPLANATIONS - MECHANICS UNLIMITED

20. Normally do not deteriorate.

32. Perform 'a' if bearing overheats.

40. You could get into trouble. Secure boilers.

42. After isolating electrical source.

50. Refers to roller bearing or shrunk-fit piece.

ANSWERS - DIESEL ENGINEER EXAM

1.a
2.b
3.c
4.a
5.a
6.b
7*c
8.c
9.b
10.b
11.a
12*c
13*c
14.d
15.d
16.a
17*a
18.a
19*b
20.c
21*c
22.c

23.c
24.c
25*d
26*b
27.b
28.a
29.a
30.c
31.a
32*c
33.c
34*c
35.a
36.c
37.c
38.c
39.d
40.a
41.a
42*c
43*a
44.a
45*c
46.a
47.d
48*b
49*a
50.c
51.d
52.b
53.c
54.b
55.c
56.c
57.c
58.a
59.d
60.d
61.a
62.a
63 b
*64*b*
*65*d*
*66*a*
67.c
68.a
69.b
70.d
71.b
*72*c*
73.d
74.a
75.a
76.b
77.a
*78*b*
79.c
*80*c*
81*b
82*a
83*c
84.b
85.b
86.d
87.c
88.a
89.d
90.a
91.d
92.a
93.c
94.b
95.b
96.c
97.b
98.c
99.b
100.d
101*c
102*a
103.a
104.b
105.b
106.b
107.d
108*d
109*b
110.a
111.c
112.a
*113*c
114.d
115.a
116*a
117.a
*118*e
119.a
120.b
121.b
122.b
123.c
124.a
125*e
126.d
127.e
128*d
129.d
130.c

131.a
132.a
133*a
134.a
135*b
136.e
137.b
138*e
139*b
140.c
141.b
142*c
143.a
144.b
145.a
146.a
147.b
148.c
149*a
150.b
151*a
152.a
153.a
154*a
155*a
156.d
157.a
158.d
159.a
160*c
161.b
162.a
163*d
164.d
165.b
166.a
167.b
168.a
169.b
170.d
171.d
172.d
173.a
174.a
175*a
176.b
177.d
178.c
179.d
180.b
181.a
182.b
183.c
184*a
185.c
186*b
187.a
188.a
189.d
190.a
191*c
192.b
193*c
194*c
195.a
196*a
197.a
198.c
199*b
200.a
201.d
202.c
203.a
204.d
205.d
206.d
207.b
208.b
209*
#1.b
#2.b
210.b
211*b
212.a
213.b
214*c
215*d
216.c
217.b
218*c
219.d
220*b
221.c
222.e
223*c
224*c
225.a
226.d
227.b
228.c
229.a
230*b
231*c
232.b
233.c
234*c
235*a
236.a

237.a
238.a
239*a

EXPLANATIONS - DIESEL ENGINEER EXAM

7. Poppet valves for exhaust (not like motorcycle engines with rotary valves).

12. Flow rate = time and speed of air entering, and exhaust leaving = lowered restrictions.

13. Lets air into cylinder and gaseous fuel at the same time.

17. 2 stroke or 4 stroke with overhead cam.

19. Also stronger.

21. Cylinder cooling by water jackets. Scavenging also cools cylinder in the process.

25. Ports are the air entry openings or outlets.

26. 2 stroke with exhaust valves or 4 stroke engine.

32. Exhaust pressure and velocity is high enough to operate turbine.

34. Called, ‘valve overlap’.

42. 4 stroke and 2 stroke have cams in diesel engines.

43. Air expanding in turbine to cool air and assist turbocharger drive output.

45. Requires special valve gear to accomplish this. Expanding any gas lowers gas temperature.

48. Scavenging cools cylinders.

49. Piston on TDC = top dead center.

64. Reciprocating are usually plunger type or rotary gear type.

65. Pilot oil ignition ignites gas charge.

66. Lubricating oil in cylinder will retard combustion in answer ‘c’.

72. Measured by an engine diagram chart.

78. ‘c’ is also correct.

80. In this order. Large diesels have crossheads to connect piston rod and connecting rod similar to steam engines.

81. Driven by crankshaft.

82. Tricky, but true! Gas diesels tend to detonate when running on full diesel effects.

83. Except for a gas diesel ‘a’ would be correct.

101. Oil tank may be in any location.

102. Fuel oil transfer pumps are ‘c’.

104. One to transfer to day tank, another to inject to cylinders.

108. I hope you didn't mark answer ‘c’.

109. A spring loaded relief valve opens if filters clog to prevent loss of oil to bearings.

113. Piston at TDC.

116. Valve overlap taking place.

118. Some are only on ‘b’.

125. ‘c’ raises compression pressures above normal.

128. ‘a’ will cause overheating as oil must circulate to take heat from injector

133. "Require" should be, "result in." Overloading engine will also do this.

135. Unless turbos are tied to a common discharge perform ‘a’.

138. Ball joint instead of a wrist pin is used so piston can rotate.

139. Take note of ‘auxiliary line’. Crosshead engines employ these.

142. Overloading will cause this. Detonation if severe, but wear is shown on bearings.

149. At reduced load of course. Engine must have piston scavenging effects to do this.

151. Usually in cylinder heads.

154. Compression and combustion temperatures are very high.

155. Cam which controls unit injectors.

160. If oil pump was shaft driven, it couldn't pump in reverse and supply bearings with oil.

163. Blades will always have deposits. A corroded blade is considered defective.

175. To balance combustion pressures even the load among cylinders.

184. If no flow regulators are installed on cooling lines.

186. 'a' will happen partially, but rotation means complete rotation and this engine will not run in reverse so 'b' will happen.

191. Shaft driven oil pump is normally not in use on turning gear as auxiliary pump is used. If shaft pump is used 'b' will occur.

193. Or 'Hertz' (Hz) when using synchronizing switch to check generator. Both are in relation to each other 3,600 r.p.m.'s = 60 Hz in most applications.

194. Steam condensing from exhaust.

195. 'c' is very undesirable, causes detonation.
196. Smoke is wrong terminology for vapor. Usually oil will drip from breather crankcase vent as oil floats on water being displaced by water. 'b' does happen, but difficult to detect. Governors have their own oil supply.

199. Higher temperature to ignite fuel and is used in low speed engines.

209. To avoid accidental starting.

211. 'c' if governor is oil energized.

214. A lean mixture. Smoke is a proper term.

218. 'c' is something new to learn and is correct. Very catchy on an exam.

220. Spark plug or pilot oil.

223. Oxygen and oil catches fire on contact.

224. Large diesels use cooling towers. Sulfite is used in closed systems and absorbs O_2 . Why don't they use sulfite in question #224?

230. 'a' at cam overlap and 'b' with exhaust valves with a blower is correct. 'b' is chosen because piston is at lowest point exposing "cylinder" during admission.

231. Misfiring does not refer to backfiring. Good to keep in mind.

234. Injectors must exceed compression pressure to spray fuel unrestricted.

235. Diagram shows engine pressures and valve timing.

239. Water or oil will not be compressed. Cylinder relief should open to relieve, but will not handle large volume.

240. Inferior fuel also will detonate causing cylinder and piston knocks. Loose bearings will also knock.

242. Slightly, if dissolved gasses are present.

243. At worse it will throw crank out of balance or break it.

253. Always use manufacturers specifications.

261. 'e' causes "valve float" - rocker leaves cam and doesn't return to cam base circle fast enough so valve remains open. 'd' occurs.

Chapter 20

ILLUSTRATIONS

Refer to the illustrations, so you can see how the boiler plant is constructed.

Implant the "picture" in your mind, then refer to this memory when answering the questions in any written or oral examination.

GO TO NEXT PAGE

LEARN EVERY COMPONENT IN THE POWER PLANT

'STEAM & DIESEL POWER PLANT EXAMS'

... BEST-SELLING EXAM BOOK SINCE 1981 ...

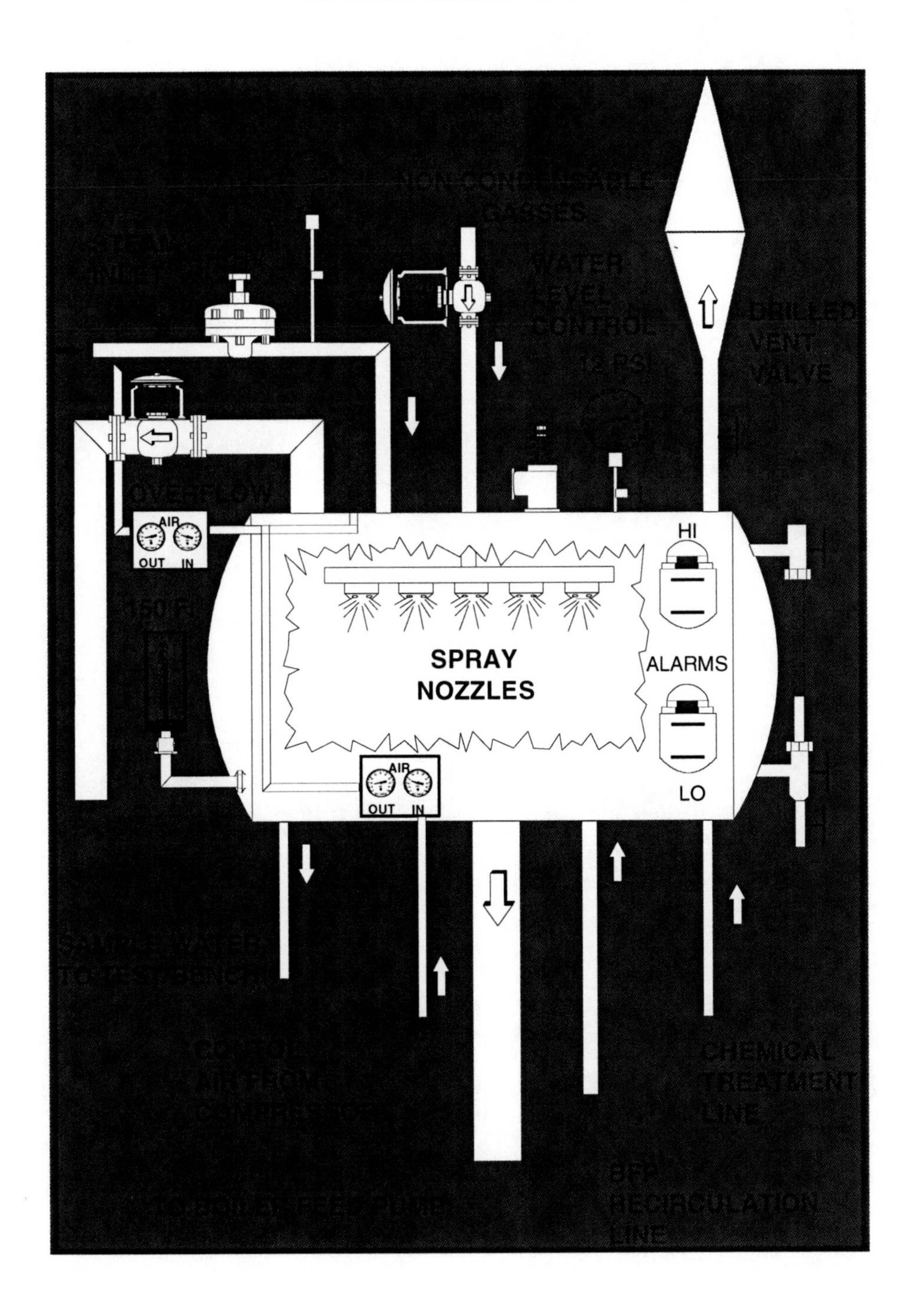

DEAREATOR

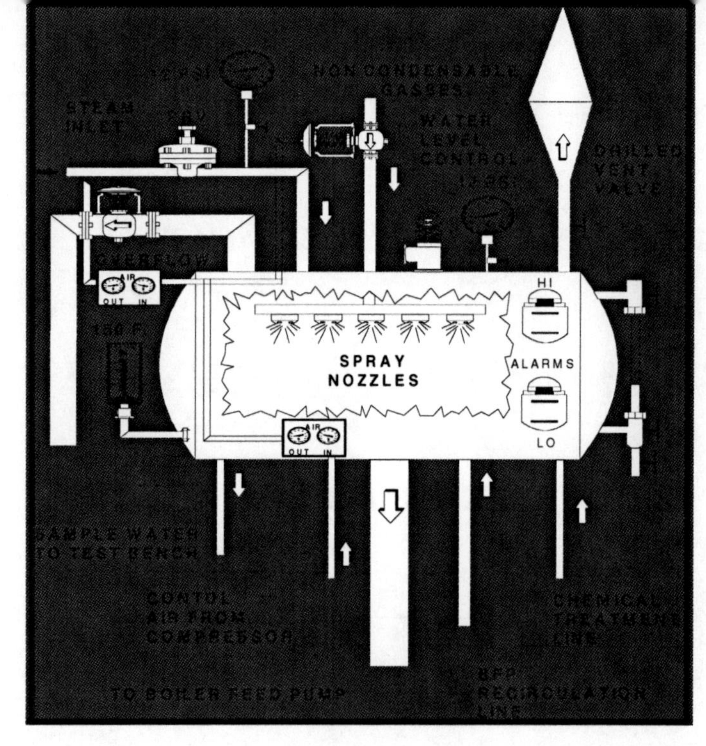

To the left is a compressed drawing of a feedwater deareator heater revealing all its controlling devices. Learn all you can about every device in the power plant. Learn what to check on your routine shift inspections and what to do when something goes wrong! You can see how you will be well prepared by breaking down the components of a power plant. Now you will see how it all fits in together when you look at the steam & water cycle diagram.

If you pass your exam and this book helped you --tell a friend. One day, that person you help today may help you tomorrow!

STEAM & WATER CYCLE

Next Page.

STEAM & WATER CYCLE

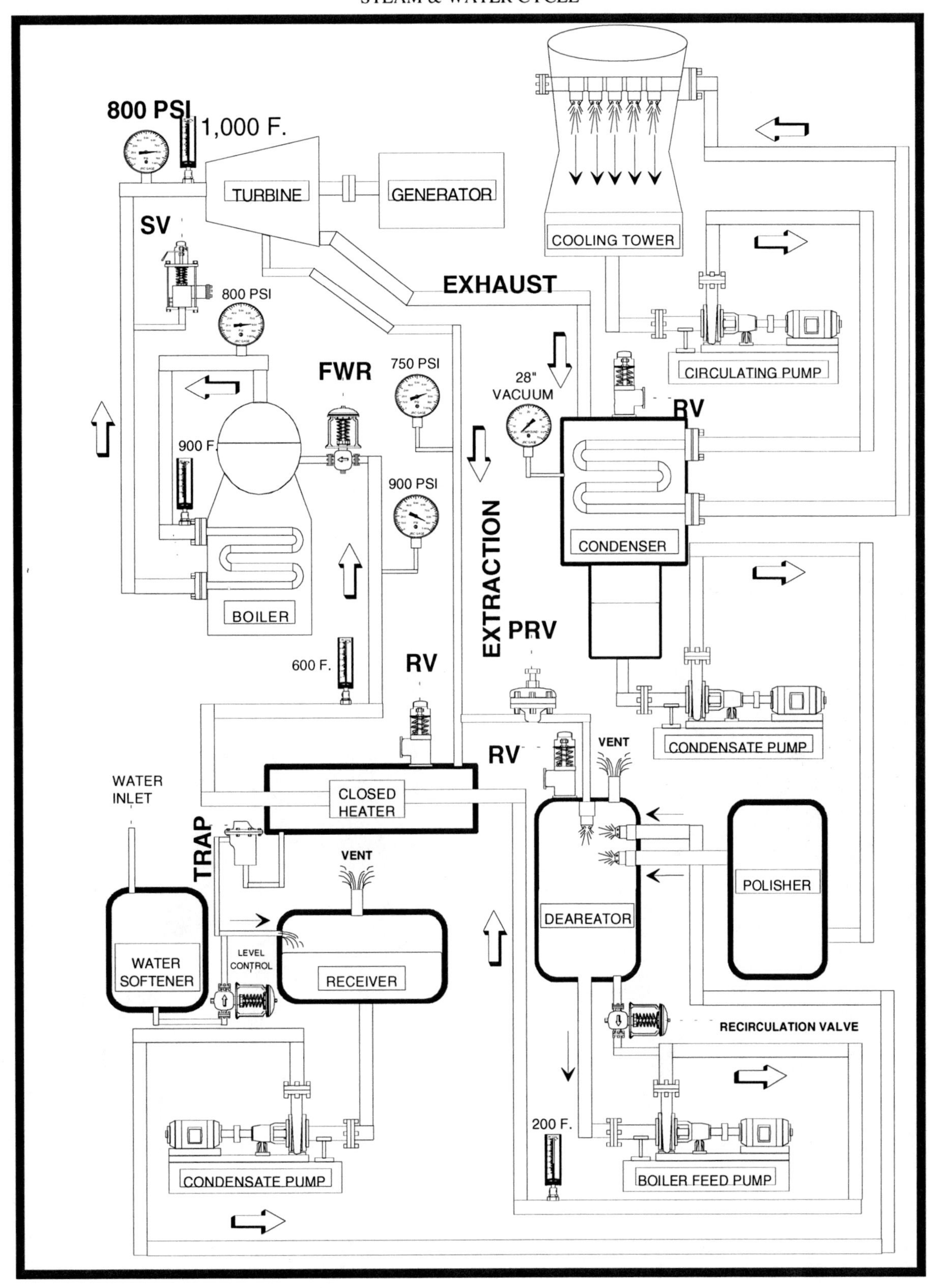

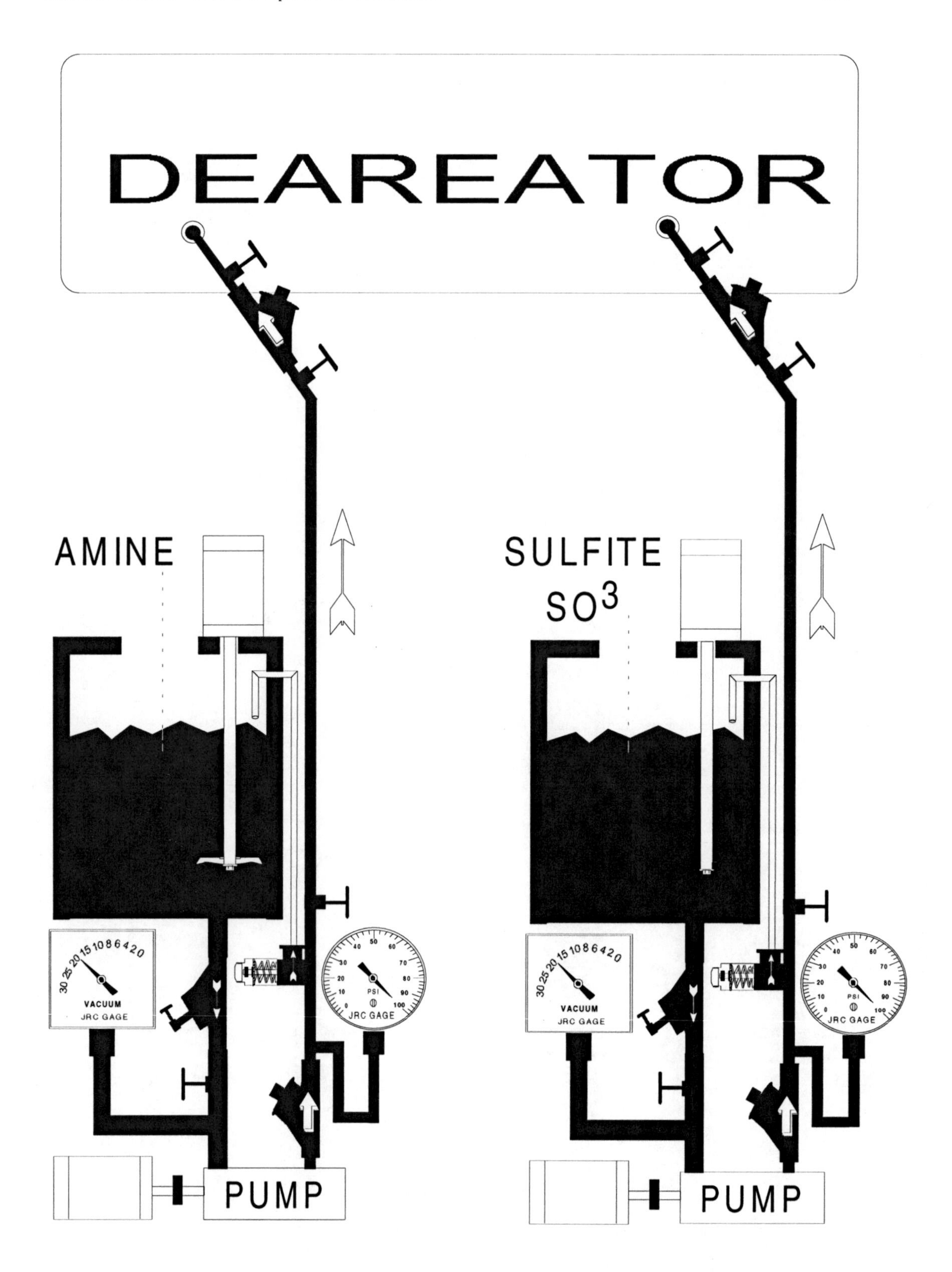

FEED WATER CHEMICAL INJECTION SYSTEM

BLOW DOWN SYSTEM (Next Page)

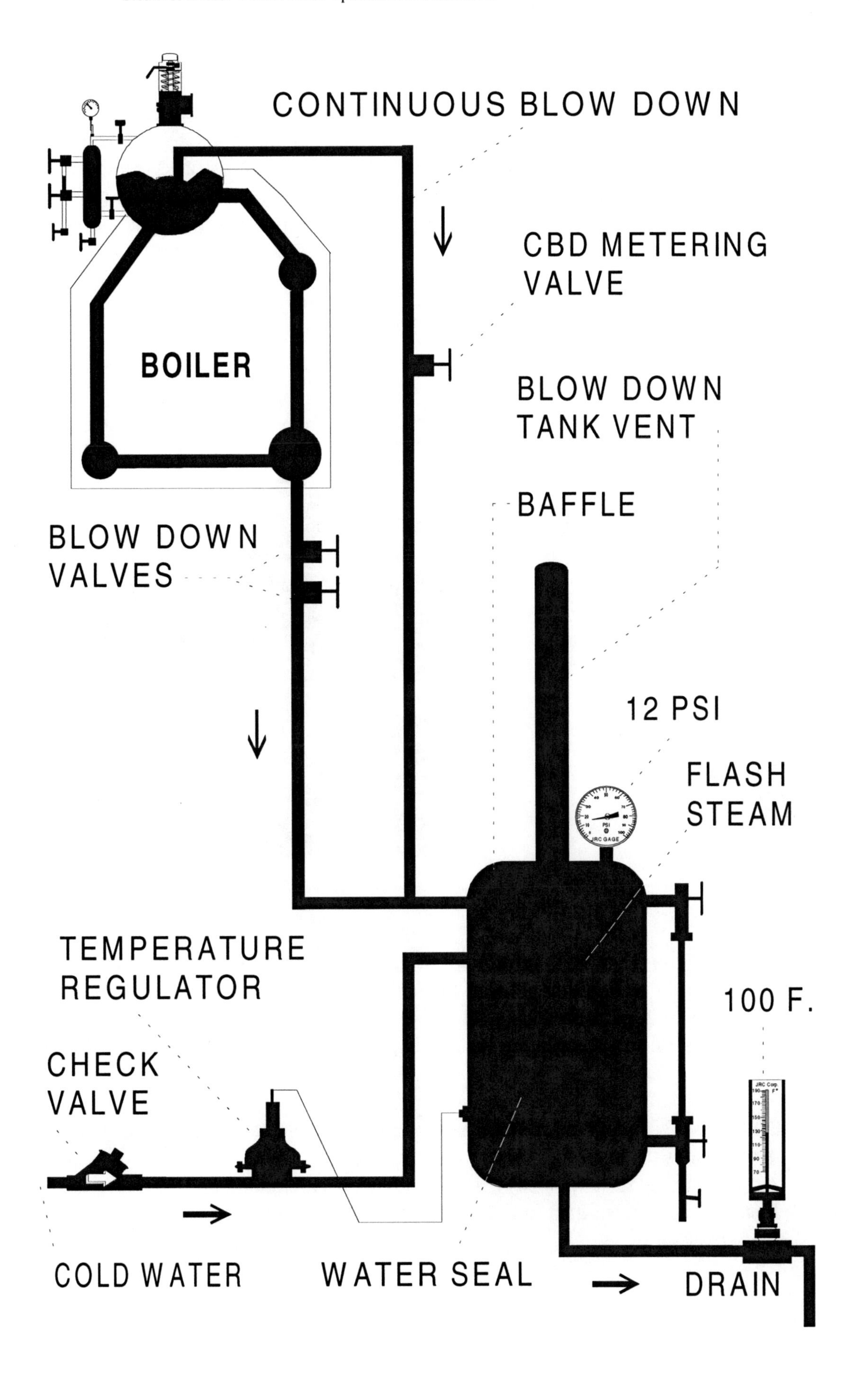
CONTINUOUS BLOW DOWN
CBD METERING VALVE
BOILER
BLOW DOWN TANK VENT
BAFFLE
BLOW DOWN VALVES
12 PSI
FLASH STEAM
TEMPERATURE REGULATOR
100 F.
CHECK VALVE
COLD WATER
WATER SEAL
DRAIN

Be careful when firing on gaseous fuels, as they are very potent. Most critical time is start up to insure the furnace is not filling with excessive fuel. Also, as shut down insure the main fuel valves are not only closed, but are not leaking. Furnace purging prior to lighting off is critical, but a purge will do no good if the fuel valves are leaking, you'll be filling the passes with gas! A simple test is to use your nose to sniff for gas. Open a peep hole or flue port and smell for gas during the purge. It could save your life!

FLAME SENSOR

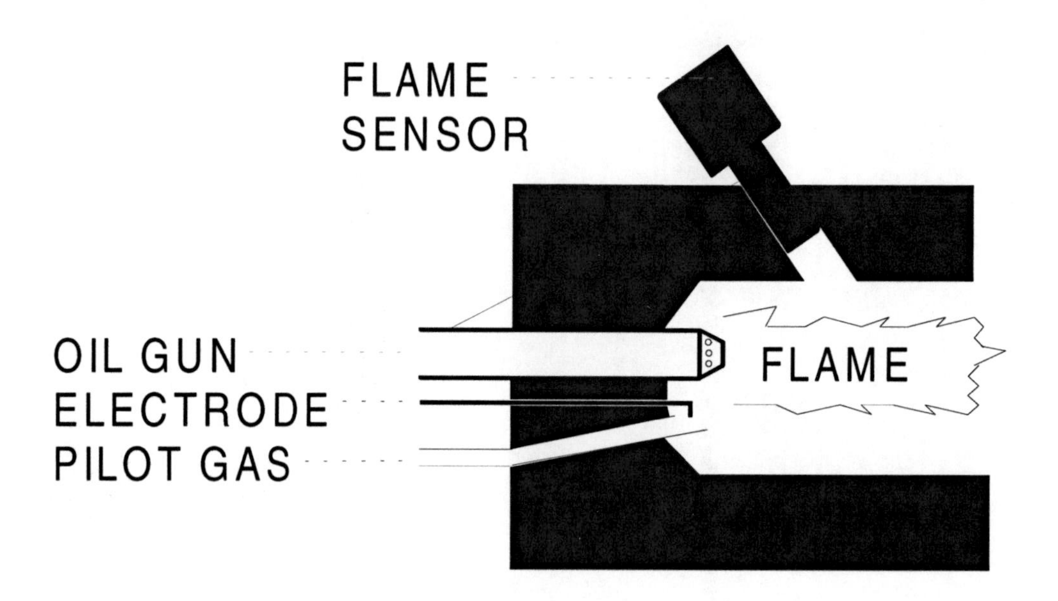

The flame sensor can be sensitive to infrared or ultraviolet light. Many old coal-fired boilers ran into a problem converting to liquid or gaseous fuels, that when the flame blew out the infrared sensor would 'see' the red-hot firebricks or oil slag on the walls and refuse to shut off the main fuel valve. This could cause a devastating furnace explosion. The ultraviolet flame sensor solved this problem.

To test a flame sensor? Simply unscrew the sensor and remove it so it can no longer see the flame. In a few seconds the main fuel valve should trip closed. Do not use a flashlight beam to counteract the safety flame sensor, unless it's a dire emergency situation and you have to keep the fires lit. In this case, be very vigilant to flame conditions as you will now be the only safeguard against a furnace explosion.

Keep in mind many safety devices are not required by codes; low water cut-out, flame sensor, high and low fuel oil temperature alarms, etc. As the boiler operator, you are the ultimate safety device, so pay attention on the job and do not rely on safety devices as they can fail!

TO PASS LICENSING AND EMPLOYMENT EXAMS

Take this book with you to any high pressure boiler plant and ask the Chief Engineer if you can stay on a few shift watches to learn. Most will accommodate you! It won't take long to learn as you go over each question & answer in this book. Memorization will be eliminated to a great degree as you see the operation procedures being performed. Now, when any power plant questions is asked, you'll be able to see in your mind's eye the correct answer and procedure. It can take a little as three weeks!

It is not very difficult to pass written and oral exams. The trick is to be exposed to the equipment so you become familiar with its operation. Then, with this book, take the exams when you are inside the power plant, so you can walk up to the device and grasps a mental picture of the object related to the question.

I have done this and have never failed any power plant exam; written or oral, all the way up to the unlimited classes of licensing! Using this simple technique makes taking exams easy. Why? Because you are not relying on memory, but experience. You can't memorize every word in this book, but you can photograph in your mind every component within the power plant. Go ahead, close your eyes and you will see that you can visualize the feed pumps, boilers, valve locations, even a picture hanging on the wall. Get the idea?

BOILING GOSSIP

You may not think this to be important, but it really is. Don't get involved with boiler room gossip. Plant operators are often like a bunch of teenage girls gossiping like they were in high school. Look, everybody is incompetent, because nobody can be competent all of the time, so mistakes are made. Some are promoted and others feel the person does not deserve it. So what? Can gossip demote the person? Of course not. In fact, gossip changes nothing but igniting hot coals on your own head. Life is short, so don't waste time with destructive gossip. The sooner you do, the more happier life you will have. It may even save your job!

SAFETY FIRST

Don't play around with power plant equipment if you are uncertain of its operation. Get training! Power plants are not safe places to work in. There are so many hazards to be scalded, slip and fall, electrocution, suffocation, loss of limb, you name it. To be safe serves to save your own life and those who work with you.

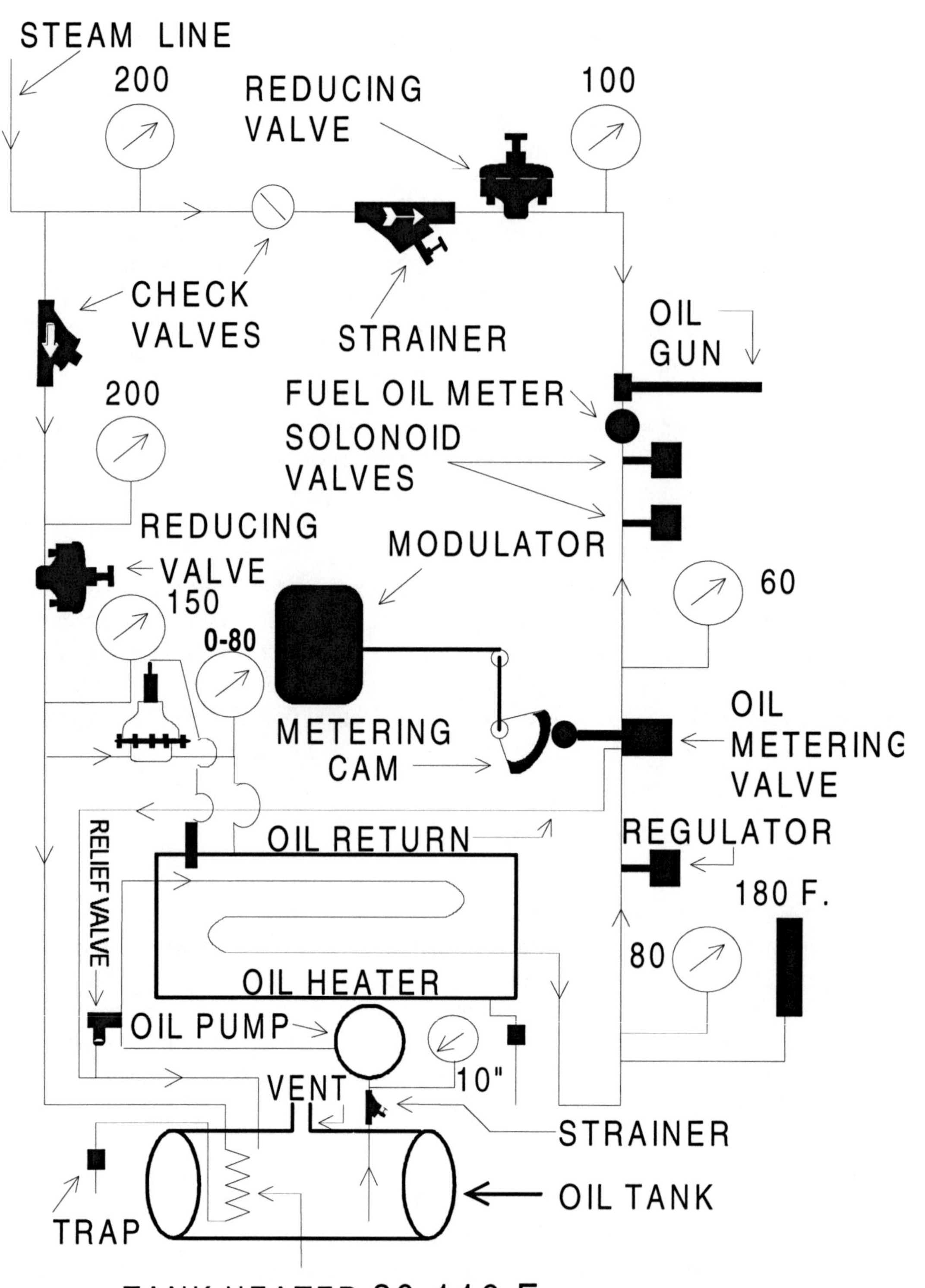

FUEL OIL CYCLE

JAMES RUSSELL PUBLISHING

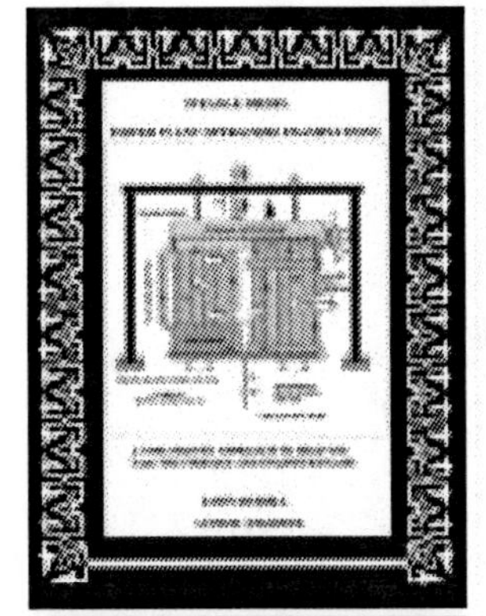

STEAM & DIESEL POWER PLANT OPERATORS EXAMINATIONS
ISBN 0-916367-088 117 pp., 8x11, illustrated, $34.95 Over 1,400 multiple-choice test questions & answers (with explanations) helps stationary engineer power plant operators pass steam boiler licensing and pre-employment exams. No other book compares with the sheer power to educate employees in the safe operation of plant equipment. A bestseller to the power plant industry since 1981. Questions from this book are used in city & state licensing and civil service exams!

TRAP SHOOTING SECRETS
ISBN 0-916367-096 183 pp., 8x11, 85 illustrations, $34.95. There has never been a book like this, ever! *TSS* is like having a shooting coach telling you precisely what to do to hit the targets. It is the first book ever to be endorsed by professional trapshooters! Hall of Fame and Olympic target shooters recommend this book! TSS transforms shooters into winning champions! All shotgun sports shooters benefit from these powerful instructions.

PRECISION SHOOTING - THE TRAPSHOOTER'S BIBLE
ISBN 0-916367-10X 117 pp., 8x11, 145 illustrations, $34.95. The *only* trap shooting book with ATA & Olympic Double-Trap technical instructions. The *only* professional advanced-level trapshooting book in the world. This target shooting book includes 315 questions & answers to master shooting proficiency. Readers obtain high-score results! Endorsed by professional shotgun shooting instructors. Winning advice for all shotgun sport shooters!

SCREEN & STAGE MARKETING SECRETS
ISBN 0-916367-118 177pp., 8x11, 60 illustrations, $34.95. This is the *only* book specifically written for writers to sell screenplays and stage plays to literary agents and the movie industry. Script formats, sample query letters, marketing instructions, contacting producers, stars, studios and even a valuable list of WGA agents willing to give readers of this book special consideration! No book compares with the marketing power of this book. Writer's learn how to sell their scripts!

ORDER YOUR BOOKS TODAY!

PUBLISHER DISCOUNTS
Bookstre Special Order 20%
Individuals $34.95
S & H $4 single copy.
E-mail:
screenplay@powernet.net

PUBLISHER - SAN 295-852X
James Russell Publishing
780 Diogenes Drive, Reno, NV 89512
Phone / Fax: (775)348-8711
www.powernet.net/~scrnplay

DISTRIBUTORS
Baker & Taylor (800)775-1100
www.btol.com
Ingrams (800)937-8000
www.ingrambook.com

THE 7 DAY PLAN TO BE A BETTER CHRISTIAN!

SUNDAY -- This is a day of rest (see Saturday) of which no work is to be performed. Take full advantage of it! However, extend extra kindness to others. Read the Word, listen to Christian radio and watch TV for faith comes by "hearing" the Word of God.

MONDAY -- Drive your vehicle with patience towards others. Be changed at work. No more gossip, complaining, bad jokes. Just start being nice -- Biblically correct! Be cooperative. Can you do this for just one day?

TUESDAY -- Forget Me! Do a good thing for another. Open doors, buy someone a meal or gift, feed a stranger's parking meter. Give so you will receive. Give something! The Lord gives, so should you.

WEDNESDAY -- Compliment Day! Say something nice to someone, including one who may not like you. Be sincere about it! If someone needs help, go to their aid. Make someone smile today!

THURSDAY -- Distribute a Bible track. No tracks? Make or buy some! It is time you begin your ministry to the Lord to share the Good News. There are many hurting people who need the Lord and it is your responsibility to introduce them to Him. Using tracks make the job easy!

FRIDAY -- Day of forgiveness! When you forgive others transgressions, you are released from the anguish within yourself. It is easy to do! Start the process today! See Tuesday and Wednesday's instructions. Life is so much easier to live and great mercy and blessing arrive when you forgive!

SATURDAY -- Rest if this is the Sabbath you honor or donate; time, items, food, or money to the homeless shelters. Do not forget the poor! Visit or call a relative or friend. Express your appreciation for what the Lord has given you! Share with others what you have and the Lord will give you even more!

Free Bible Tracks For SASE!

EACH DAY

Contact Us For Bible Tracks!

START the day right by greeting the Lord and giving thanks for all He has done and what He will do for you in the future. **END** the day right by expressing your gratitude to the Lord.

SPEAK often to the Lord, as he is your best friend. Remember, he wants to handle every detail in your life, even the small stuff. Do not become so busy in your day you leave Him out of your life.

WHEN you pray just speak as you would to a friend. There is no need for theatrical displays of emotions or insincerity. If you fall short, do not turn your face away from the Lord and hide. Take the issue to Him.

WHAT will you give to the Lord if He grants your request? Will you simply say thank you and forget Him until you need something else later? The Lord sees the suffering of the sick and poor. Why not pledge to help them? Make your promise and keep it! Do it now before you recieve. This is faith in action.

SPREAD the Word of God. You may not be a minister, but you can distribute tracks. Leave them everywhereyou go. Keep some on your person each day. Your reward shall be great! Write us for tracts!

TITHE to the Lord. Give and you shall recieve more! Give to churches, ministries, homeless shelters, or where there is dire need. A perfect expression of love for others! God's System Never Fails!

PRINT AND DISTRIBUTE TO OTHERS!

Printed in the United States
72515LV00003B/325-344

9 780916 367084